Introducing Philosophy of Mathematics

Introducing Philosophy
of Mathematics

Michèle Friend

ACUMEN

*Dedicated to my parents, Henriette and Tony Friend.
I wish that their nobility of spirit were more commonplace.*

© Michèle Friend 2007

This book is copyright under the Berne Convention.
No reproduction without permission.
All rights reserved.

First published in 2007 by Acumen

Acumen Publishing Limited
Stocksfield Hall
Stocksfield
NE43 7TN
www.acumenpublishing.co.uk

ISBN: 978-1-84465-060-6 (hardcover)
ISBN: 978-1-84465-061-3 (paperback)

British Library Cataloguing-in-Publication Data
A catalogue record for this book is available from the British Library.

Designed and typeset in Warnock Pro by Kate Williams, Swansea.
Printed and bound by Cromwell Press, Trowbridge.

Contents

Acknowledgements	vii
Preface	ix

1. Infinity — 1
   1. Introduction — 1
   2. Zeno's paradoxes — 2
   3. Potential versus actual infinity — 7
   4. The ordinal notion of infinity — 12
   5. The cardinal notion of infinity — 13
   6. Summary — 22

2. Mathematical Platonism and realism — 23
   1. Introduction — 23
   2. Historical origins — 23
   3. Realism in general — 26
   4. Kurt Gödel — 35
   5. Penelope Maddy — 37
   6. General problems with set-theoretic realism — 41
   7. Conclusion — 46
   8. Summary — 47

3. Logicism — 49
   1. Introduction — 49
   2. Frege's logicism: technical accomplishments — 52
   3. Frege's logicism: philosophical accomplishments — 58
   4. Problems with Frege's logicism — 63
   5. Whitehead and Russell's logicism — 66
   6. Philosophically, what is wrong with Whitehead and Russell's type theory? — 71

	7. Other attempts at logicism	78
	8. Conclusion	78
	9. Summary	79

4. Structuralism — 81

1. Introduction — 81
2. The motivation for structuralism: Benacerraf's puzzle — 83
3. The philosophy of structuralism: Hellman — 85
4. The philosophy of structuralism: Resnik and Shapiro — 90
5. Critique — 96
6. Summary — 100

5. Constructivism — 101

1. Introduction — 101
2. Intuitionist logic — 106
3. *Prima facie* motivations for constructivism — 113
4. Deeper motivations for constructivism — 114
5. The semantics of intuitionist logic: Dummett — 121
6. Problems with constructivism — 123
7. Summary — 124

6. A *pot-pourri* of philosophies of mathematics — 127

1. Introduction — 127
2. Empiricism and naturalism — 130
3. Fictionalism — 134
4. Psychologism — 137
5. Husserl — 141
6. Formalism — 147
7. Hilbert — 153
8. Meinongian Philosophy of Mathematics — 157
9. Lakatos — 163

Appendix: Proof: ex falso quod libet — 167
Glossary — 169
Notes — 177
Guide to further reading — 191
Bibliography — 195
Index — 201

Acknowledgements

I should like to thank John Shand for suggesting that I write this book, and for initial encouragement, and I should like to thank Steven Gerrard at Acumen for endorsing the proposal and publishing it.

I received very helpful and careful comments from my two reviewers, Stewart Shapiro and Alan Baker. Any mistakes that remain are my fault entirely. I should also like to thank an anonymous reviewer for helpful comments. Some colleagues have helped with the section on Husserl. These were Alena Vencovska, Jairo DaSilva and Marika Hadzipetros. I should also like to thank Graham Priest for comments on the paper that underpins the section on Meinongian philosophy of mathematics, and for the many audiences to whom I have exposed papers that underscore some of the other sections. These include the philosophy departments at the University of Hertfordshire and George Washington University, the mathematics department at George Washington University and particularly the audience for the Logica '05 conference in the Czech Republic. I should like to thank David Backer for helping with the final notes and bibliography. Kate Williams edited the text and produced the illustrations.

On a more personal front, I should like to give special thanks to my parents, my husband, my enthusiastic seminar students and my friends, who encouraged me to write, although they knew not what. I should also like to thank the philosophy department at George Washington University for academic, personal and financial support.

Michèle Friend

Preface

This book is intended as an upper-level undergraduate text or a lower-level graduate text for students of the philosophy of mathematics. In many ways the approach taken is standard. Subjects discussed include Platonism, logicism, constructivism, formalism and structuralism; others that are less often discussed are also given a hearing.

This is not meant to be a comprehensive handbook or definitive exhaustive treatment of all, or even any, of the ideas in the philosophy of mathematics. Rather, this book contains a selected set of topics that are aired in such a way as to give the student the confidence to read further in the literature. A guide to further reading is given at the back of the book. All the books cited are in English, and should be available from good university libraries. Having read this book, the student should be equipped with standard questions to bear in mind when doing further reading. The arguments rehearsed in the text are by no means the final word on the issues. Many open questions reveal themselves, inviting further investigation. Inevitably, some of my prejudices can be detected in the text.

Most of the chapters are self-contained. Anomalous in this respect are Chapter 1 on infinity and Chapter 2 on Platonism. Chapter 1 is a technical chapter. I believe that students of the philosophy of mathematics should have a grasp of what the mathematician means by "infinity", since many of the philosophies of mathematics either have something direct to say about it, or use the concept implicitly. It is also an engaging technical topic and, thereby, an interesting point of comparison between the different theories. Pedagogically, it makes sense to discuss some technical issues while the student is fresh to the work. Having worked through some technical material, the student will have the courage to tackle some more technical aspects of the philosophy of mathematics on her own. The remaining chapters are less technical, but be warned: serious readings in the philosophy of mathematics rarely shy away from discussing quite technical notions, so a good grounding is essential to further study. For example, it is usual to be well versed in set theory and model theory.

Platonism is the "base philosophical theory" behind, or acts as a point of reference for, many of the philosophies of mathematics. Most philosophies of mathematics were developed as a reaction to it. Some find that the body of mathematical results do not support Platonism; others find that there are deep philosophical flaws inherent in the philosophy. Most of the subsequent chapters refer back to infinity, Platonism or both. Cross-referencing between the subsequent chapters is kept to a minimum.

In Chapter 3 we discuss logicism, which is seen as an interesting departure from some aspects of Platonism. Usually a logicist is a realist about the ontology of mathematics, but tries to give an epistemological foundation to mathematics grounded in logic. In Chapter 4 we then look at the more recent arguments of structuralism, which can be construed as a type of realism, but cleverly avoids many of the pitfalls associated with more traditional forms of realism or Platonism.

Constructivism, discussed in Chapter 5, is a sharp reaction to Platonism, and in this respect also rejects logicism. This time, the emphasis is on both epistemology and ontology. The constructivist revises both of these aspects of the Platonist philosophy. The term "constructivism" covers a number of different philosophies of mathematics and logic. Only a selected few will be discussed. The constructivist positions are closely tied to an underlying logic that governs the notion of proof in mathematics. For this reason, certain technical matters are explored. Inevitably, some students will find that their previous exposure to logic used different notation, but I hope that the notation used here is clearly explained. Its selection reflects the further reading that the student is encouraged to pursue. Again, the hope is that by reading this chapter the student will gain the confidence to explore further, and, duly equipped, will not find all of the literature too specialized and opaque. Note that by studying constructivism after structuralism we are departing from the historical development of the philosophy of mathematics. However, this makes better conceptual sense; since we are anchoring our exploration of the philosophical approaches in infinity and Platonism. Structuralism is closer to Platonism than is constructivism, so we look at structuralism before constructivism.

Finally, Chapter 6 looks at a number of more esoteric and neglected ideas. Unfortunately, some of the relevant literature is difficult to find. Nevertheless, the chapter should give the reader a sense of the breadth of research being carried out in the philosophy of mathematics, and expose the student to lesser-known approaches that he might find appealing. This should encourage creativity in developing new ideas and in making contributions to the subject. The reader may think that many of the sections in Chapter 6 warrant a whole chapter to themselves, but by the time they have reached Chapter 6, some terms and concepts will be familiar (for example, the distinction between an analytic truth and a synthetic truth does not need explaining again), thus the brevity of the sections is partly due to the order of presentation.

There are several glaring omissions in this book, noticeably Wittgenstein's philosophy of mathematics. By way of excuse I can say that this is not meant as an encyclopaedia of the philosophy of mathematics, but only an introduction, so it is not intended to cover all philosophies. Nevertheless, the omission of Wittgenstein's philosophy of mathematics bears further justification. I am no expert on Wittgenstein, and I am not sure I would trust second-hand sources, since many disagree with each other profoundly. I do not have the expertise to favour one interpretation over others, so I leave this to my more able colleagues.

It is hoped that the book manages to strike a balance between conceptual accessibility and correct representation of the issues in the philosophy of mathematics. In the end, this introduction should not sway the reader towards one position or another. It should awaken curiosity and equip the reader with tools for further research; the student should acquire the courage, resources and curiosity to challenge existing viewpoints. I hope that the more esoteric positions having been introduced, students and researchers will take up the standards, and march on to develop them further. As we should let a potentially infinite number of flowers bloom in mathematics, we should also welcome a greater number of well-developed positions in the philosophy of mathematics. Each contributes to our deeper understanding of mathematics and of our own favoured philosophical theories.

Chapter 1
Infinity

1. Introduction

In this chapter mature philosophical ideas concerning mathematics will not be discussed in any depth. Instead we discuss various conceptions of infinity, setting the stage for more technical discussions because each philosophy has strong or interesting views concerning infinity. We had better know something about infinity before we embark on philosophical disputes. The disputes are strong. Some philosophers endorse the whole classical theory of inifinity. Others wholly reject the classical theory, finding it misguided and dangerous, and replace it with a more modest conception of infinity, or strict finitism. Friends of classical infinity include realist positions such as Platonism, logicism and structuralism; enemies include constructivism, empiricism and naturalism. Some philosophers are ambivalent about infinity. These include David Hilbert (1862–1943) and Edmund Husserl (1859–1938).

Note, however, that while many philosophies of mathematics can be cast in terms of their views on infinity, this is not necessarily the most historically, or even philosophically, accurate way of characterizing them. There are good reasons sometimes to think of disputes as revolving around other topics.[1] When this is the case, alternative axes of dispute will be carefully considered. Nevertheless, attitudes to infinity will be discussed under each philosophical position discussed in the chapters that follow. Infinity is an important concept in mathematics. It has captured the imaginations of philosophers and mathematicians for centuries, and is a good starting-point for generating philosophical controversy.

This chapter is divided into five sections. Section 2 is largely motivational and historical. It introduces Zeno's paradoxes of motion, which will unsettle any preconceived idea that infinity is a simple topic. Zeno's paradoxes were well known in ancient Greece, and attempts were made to solve them even then. From these attempts[2] the ancient Greeks developed two conflicting views: "potential infinity", championed by Aristotle, and "actual infinity", championed by followers of Plato. These two views on infinity will be discussed in §3. They immediately serve as intuitive motivators for two rival philosophical positions: constructivism and realism, respectively. Potential infinity and actual infinity are not philosophical viewpoints; they are merely ideas about infinity that partly motivate philosophical positions.

The rest of the chapter will develop the classical theory of actual infinity, since this is mathematically more elaborate. Section 4 concerns infinite ordinals. Section 5 discusses infinite cardinals and runs through Georg Cantor's diagonal argument. This introduces the student to "Cantor's paradise", which enthralled Hilbert, despite his insistence on the *practice* of mathematics being finite. Complete understanding of the minutiae of these sections is unnecessary in terms of understanding the rest of the book. It is enough if the student appreciates the distinction between ordinal and cardinal numbers, and understands that there are many infinite numbers under the conception of actual infinity. To understand this is important because the infinite numbers are considered to be part of classical mathematics, which, in turn, is underpinned by classical logic (which is what undergraduates are usually taught in early courses in logic). Classical logic is appropriated by the realists, who take it to be the best formal expression of the logic underlying mathematics.

2. Zeno's paradoxes

Notions of infinity have been around for a long time. In the ancient Mesopotamian *Gilgamesh Epic*[3] we see a concept of infinity already surfacing in the mythology: "The Gods alone are the ones who live forever with Shamash. / As for humans, their days are numbered".[4] This early notion of infinity is that of an endless existence. For us, the puzzle is how to deal with infinity mathematically. For this we have to wait several hundred years for Zeno of Elea, who flourished around 460 BCE.

Zeno of Elea wrote one of the first detailed texts on infinity. The originals do not survive, but the ideas are recounted by Aristotle and others.[5] Zeno's famous paradoxes of infinity concern the infinite divisibility of space, and thus the very possibility of motion. The paradoxes leave us bewildered. We know the word "infinity", we use it regularly, and yet, when we examine the notion closely, we see that we do not have a clear grasp of the term.

The setting for Zeno's discussion of infinity is a discourse on the paradoxes of motion, and there is both a modest conclusion and an ambitious one. The modest conclusion to be drawn by the readers or listeners was that the concept of infinity held by the leading scholars of the day was confused. More ambitiously, and dubiously, the readers were to conclude that motion, and change, are really illusory, and that only the Unchanging, or the One, is real. Modern interpreters attribute this further conclusion to the fact that Zeno was a loyal student of Parmenides, and Parmenides supported the doctrine that there is an underlying unity to the world that is essential to it. More importantly, the One is the true reality. Therefore, change is essentially illusory. Thus, we can interpret Zeno's work as supplying further evidence for Parmenides' idea that there is only the One/the Unchanging. Modern readers tend to resist this further conclusion, and certainly will not accept it simply on the basis of Zeno's paradoxes.

Nevertheless, Zeno's paradoxes are still troubling to the modern reader, who might accept the more modest conclusion that we are confused about infinity while rejecting the further claim that change is illusory. Most people today do not feel confused for long, because they think that inventing calculus was a solution to the paradoxes. However, calculus does not solve the puzzle; rather, it ignores it by finding a technical way of getting results, or by bypassing the conceptual problem. Thus, mathematicians and engineers have no problems with infinitesimals, but as philosophers we are left with the mystery of understanding them.[6]

Let us survey three of Zeno's paradoxes. The first paradox, as reported by Aristotle, is the paradox of the race course. It is argued that for a runner such as Achilles to run a race, he has first to run half the distance to the finish line (Fig. 1). Before he can run the second half, he has to run the next quarter distance (i.e. the third quarter of the race track) (Fig. 2). Before he can finish, the runner has to complete the next eighth distance (i.e. the seventh eighth) of the course (Fig. 3), and so on *ad infinitum*. Since the runner has to complete an infinite number of tasks (covering ever smaller distances) before he can finish the race, and completing an infinite number of tasks is impossible, he can never finish the race. As a flourish on the first paradox, we can invert it. Notice that before the runner runs the first half of the course, he has to have run the first quarter of the course. Before the runner runs the first quarter, he has to have run the first eighth, and so on *ad infinitum*. Therefore, it is impossible to start the race!

It is worth dwelling on these paradoxes a little. Think about any motion. For something to move it has to cross space. On the one hand, we do manage to move from one place to another. Moreover, in general, this is not difficult. On the other hand, if we think of space itself, we can divide any space, or distance, in half. It does not seem to matter how small the distance is. We can still, in principle, divide it in half. Or can we?

One possible solution to the above paradoxes is to think that there is a "smallest" distance. The process of dividing a distance in half has to come to an end, and this is not just because our instruments for cutting or dividing are too gross, but because space comes in discrete bits. At some point in our (idealized) dividing, we have to jump to the next smallest unit of space.

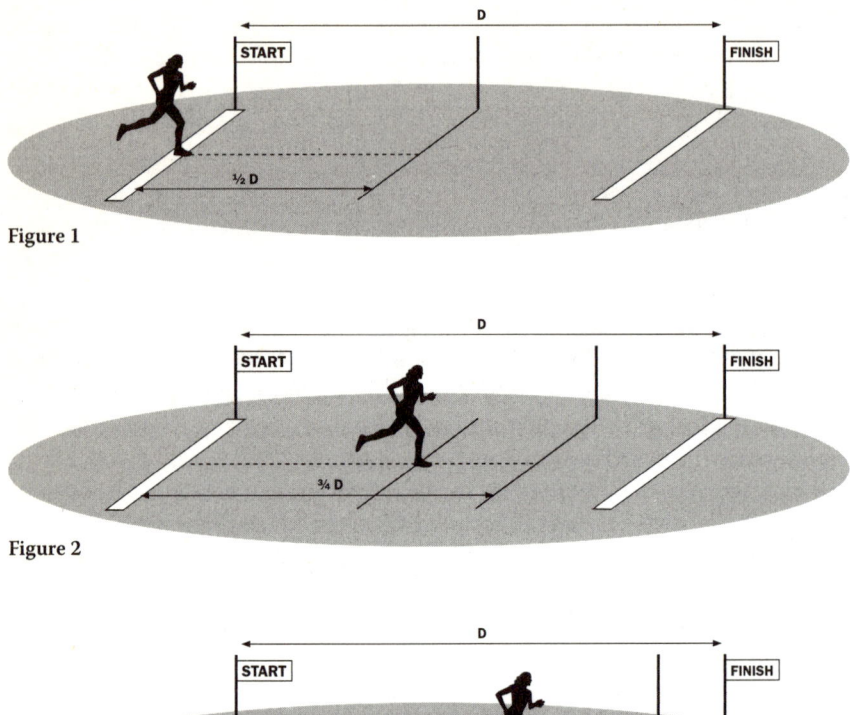

Figure 1

Figure 2

Figure 3

If space does have smallest units, which cannot be further subdivided, then we say that space is "discrete". And the same, *mutatis mutandis*, for time. Returning to our paradox, Achilles does not have an infinite number of tasks to complete in order to finish the race. He has only a finite number of smallest units of distance to traverse. This is all well and good, but we should not feel completely content with this solution because Achilles still has a very large number of tasks to complete before finishing the race. There is still some residual tension between thinking of running a race as a matter of putting one foot in front of the other as quickly as our muscles can manage, and thinking of it as completing a very large number of very small tasks. The next paradox will help us to think about these small tasks.

The second paradox is called "The Achilles" paradox,[7] or "Achilles and the tortoise". The idea is that Achilles is to run a race against the tortoise. Achilles is a good sport, so he gives the tortoise a head start by letting the tortoise

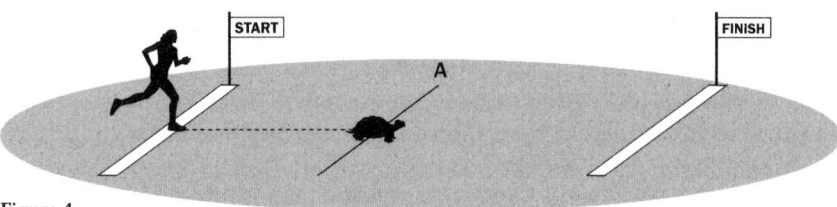
Figure 4

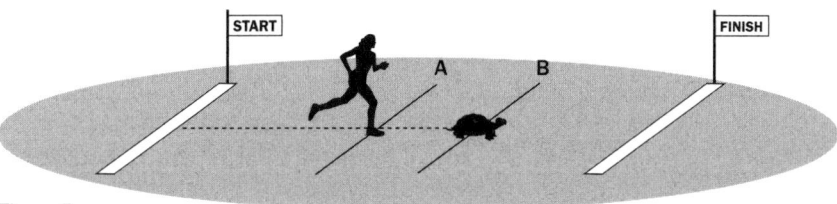
Figure 5

Figure 6

start at a distance ahead of him. They begin at the same time. Will Achilles overtake the tortoise and win the race?

As Achilles runs to catch up with the tortoise, the tortoise is also moving (Fig. 4). By the time Achilles has reached the point of departure of the tortoise, A, the tortoise will have moved ahead to a new place, B (Fig. 5). Achilles then has the task of running to B to catch up to the tortoise. However, by the time Achilles gets to B, the tortoise will have crawled ahead to place C (Fig. 6). In this manner of describing the race, Achilles can never catch up with the tortoise. The tortoise will win the race.

Notice that this is perfectly general. It does not depend on a minimal distance between the starting-point of Achilles and the tortoise, or on a particular length of race. In real life, it would make a difference. There would be starting distances between Achilles and the tortoise where the tortoise would obviously win, starting distances where Achilles would obviously win and starting distances where the result could go either way. The problem of the paradox has to do with the order of the tasks to be completed. It is quite right that Achilles has to catch up with the tortoise before overtaking the tortoise. It is also correct that the tortoise is also in motion, so is a moving target. Again, if space is discrete, then there will be a last unit of distance to cross for Achilles to be abreast of the tortoise, and then Achilles is free to overtake it. We should consider that Achilles runs faster than the tortoise, so Achilles can overcome the small distances more quickly. He can complete more tasks in a shorter time. The paradox seems to reinforce the hypothesis that space is discrete. Moreover, it makes plain that time had better be discrete too: allowing for speed to come in discrete units. Unfortunately, we cannot rest there, since there is a further paradox that is not solved by the hypothesis that space and time are discrete.

Consider the last paradox: that of the blocks. One is asked to imagine three blocks, A, B and C, of equal size and dimension. Blocks A and B are next to each other occupying their allotted spaces. Block C is in front of block A (Fig. 7). Block C might move, or be moved, to the position in front of B (Fig. 8). Let us say that this takes 4 moments. Now compare this to the situation where not only is block C moved to the right, but blocks A and B are moved to the left at the same time (Fig. 9). In this case, it takes only 2 moments for the relative positions of the blocks in Figure 8 to be assumed, so the blocks reach the same relative position in half the time.

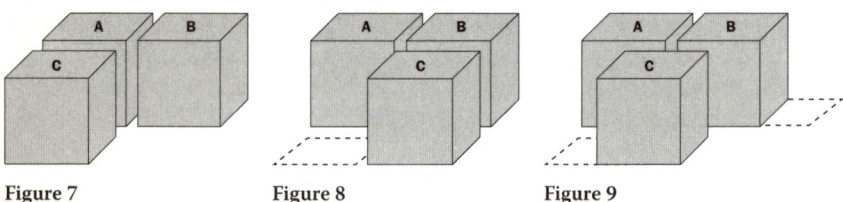

Figure 7   Figure 8   Figure 9

This is not a conceptual problem until we start to make the space occupied by the blocks maximally small, and the movements of the blocks very fast. The blocks make "moment jumps" as in the conception above. But since the blocks can move in opposite directions relative to each other, they are jumping faster relative to each other than they can relative to the ground. Again this is not a problem, except that by choosing a moving reference point we can conceptually halve the speed of a block. This tells us that speed is not only relative (to a reference point), but also can, in principle, always be halved. Space-time seems to be infinitely divisible and, therefore, not discrete.

To summarize, we can solve the first two paradoxes by arguing that space and time come in "smallest" units. If this is the case then it is false that Achilles has an infinite number of tasks to perform to finish the race. He has only a finite number of tasks: a finite number of space units to cross. This notion of space and/or time being discrete (having smallest units) solves the second paradox too. But if space and time are discrete, then we cannot solve the last paradox. For the last paradox just shows that we can always subdivide units of space and time (by changing the reference point from stationary to moving). Since, in principle, we could occupy any moving reference point we like, we should be able to infinitely subdivide space and time. Therefore, our conception of infinity bears refining. Zeno wanted his readers to conclude that motion is illusory. We do not have to accept this further conclusion.

We shall say no more about these paradoxes. There have been many good studies of them, and they are introduced here just to show that some work has to be done to give a coherent account of infinity. In particular, in resolving the paradoxes as a whole there are two conflicting ideas: the notion of space and time as always further divisible (we call this "everywhere dense") and that of space and time as discrete.[8]

This brings up another issue about infinity that was debated in the ancient world: what do we really mean by "in principle infinite"? More specifically, we need to choose between the notions of potential infinity and actual infinity. We turn to this pair of concepts in the following section.

3. Potential versus actual infinity

As a result of contemplating Zeno's paradoxes, Aristotle recognized the conceptual confusion surrounding infinity. He developed his own notion of infinity, drawing a distinction between potential infinity and actual infinity. We can think of the concept of potential infinity as "never running out, no matter what", and the concept of actual infinity as "there being (already collected) an infinite number of things: temporal units, spatial units or objects". Once Aristotle made this distinction, he decided that the notion of actual infinity was incoherent.

The notion of potential infinity is that of "not running out". For example, when we say that the numbers are potentially infinite, what we mean is that we will never run out of numbers. Similarly, when we say that time is potentially infinite what we mean is that there will always be more of it. This is not to say that each of us individually will not run out of time. Rather, the potential infinity of time concerns the structure of time itself. There is no last moment, or second; for each second, there is a further second.

Characteristic of the notion of potential infinity is the view that infinity is procedural; that is, we think of infinite processes and not of a set comprising an infinite number of objects. The notion of potential infinity is action-oriented (verb-oriented). We think of taking an infinite number of steps, of counting to infinity, of taking an infinite amount of time. We can do all of these things in principle. The point is that we have no reason to believe that we will run out, no matter how much we extend our existing powers of counting.

There is ambiguity in our expression of the notion of potential infinity. Compare the following statements.

(i) It is guaranteed that we shall never run out.
(ii) We are confident that we shall never run out.
(iii) More conservatively, until now we have not run out, and there is no reason at present to suppose that we will come to an end of the process.

Let us keep these three possibilities in mind, and contrast them to the notion of actual infinity.

Where potential infinity is procedural, actual infinity is static and object based. That is, we think of infinity in a very different way when we think of actual infinity. For example, we might think of the set of even numbers, and say that this is an infinite set. Moreover, it is an object we can manipulate; for example, we could combine it with the set of odd numbers and get the whole set of natural numbers. If we say that time itself is actually infinite, we mean this in the sense that time can be represented by a line that has no ending (possibly in both directions, or possibly only in one). The actually infinite time line being represented is an object that we can discuss as a whole. That is, when we think in terms of the actual infinite, we think of infinite objects: sets or dimensions or some other objects that we can treat as a collected whole. So the infinite object is an object: a set with an infinite number of members, parts or extension.

We can now contrast the conceptions of actual and potential infinity. Recall our three expressions of potential infinity. The first – "it is guaranteed that we shall never run out" – is somewhat odd in that we are tempted to ask what it is that guarantees that we shall never run out. The advocate of actual infinity will simply respond by explaining that the guarantee that we shall not run out comes from the existence of an infinite set, in terms of which the procedure of counting is couched. Put another way, when I say that we are guaranteed never to run out of numbers, what sanctions the guarantee is that there is a set of numbers that is infinite. Thus this expression of potential infinity relies on acceptance of the notion of actual infinity, so the two notions are not incompatible. We have the notion of a potentially infinite procedure, guaranteed to be infinite because the number of possible steps is infinite. More explicitly, the procedure sits on top of, and depends on, an actually infinite set. Under this conception, we just have to be careful about whether we are discussing infinity as a procedure or as an object, because we can do both.

Recall that Aristotle thought that the notion of the actual infinite was incoherent, so expression (i) of potential infinity is not one Aristotle would have favoured.

The second expression of potential infinity – "we are confident that we shall never run out" – is more psychological. We can place our confidence either in the existence of the actual infinite, or in past evidence. Begin with the first. If we say that what we mean when we say that, for example, "time is infinite" is that we are *confident* that time will continue, then our *confidence* resides in there being an infinite dimension called "time". This again couches potential infinity in terms of actual infinity. Our confidence is placed in the potential infinite because we are confident of the existence of an actually infinite dimension called "time". So in this case the notion of potential infinity again depends on a notion of actual infinity, as in expression (i). Thus the two conceptions are again compatible; and this does not sit well with someone convinced of the incoherence of the notion of actual infinity.

We could take another tack and deny that our confidence depends on actual infinity. We could say that our confidence is not placed in some "spooky object" called the "time dimension" but rather in past experience. In other words, "we are confident that time is infinite" just means that in the past we have not run out, and there is no reason to think that time will suddenly stop. Maybe this is because the ending of time is inconceivable, or maybe "there is no reason" just means that there has not been one in our past experience. So either we have to explain why the infinity of time depends on our powers of conception, and this is implausible because we might just lack imagination, or expression (ii) collapses into expression (iii). Unfortunately, we cannot really tell what will happen in the future. If we are honest with ourselves, we realize that whatever we take to be potentially infinite could come to an end at any moment, even if we cannot think what this would be like. In some sense that is alright, since if time came to an end we would not be thinking at all.

Now we have to be careful. Expression (iii) is compatible with there being a finite amount of, for example, time. That is, there might be an end of time. While we may have no evidence that the stopping of time is imminent, lack of evidence does not mean that time cannot simply end. It is not even clear what such evidence would "look like", and so how we would recognize such evidence if there were any. For all we know, time might just stop tomorrow, in a billion years, in many more years or not at all. The problem is that we lack evidence, based on past experience, to help us decide. Since Aristotle thinks that the notion of actual infinity is incoherent, he has to be seriously considering expression (iii) as articulating what he means by "potential infinity".

Pushing this Aristotelian position further still, let us consider four ways of making it more precise:

(a) *All* things come to an end, so time will also. It has not yet happened.
(b) *Probably*, time is potentially infinite.

(c) Time *might* be finite *or* it *might* not.
(d) We should not push this enquiry further.

Each of these sounds harmless enough, but each has some difficulties. We ask of the person who resorts to (a) what "all" means. When we say that "all things come to an end" we mean that any procedure we can think of will come to an end. Unfortunately, there does not seem to be any guarantee of this at all, at least in the world of experience. Plenty of things carry on after a person's death, so even when a particular life has ended, some things continue. This is true for everyone we have met so far, at least as far as we know. In fact, our evidence is not purely personal evidence; it is shared evidence. We cannot say that all things come to an end, because even collectively we do not experience all things ending. Maybe we can appeal to scientific theory. So maybe science tells us that everything comes to an end, as well as having had a beginning. Unfortunately, science has not ruled on this yet, at least if we are discussing the origins and ends of the universe, temporally or spatially. The only law in physics telling us that "everything comes to an end" is the second law of thermodynamics, the law of entropy, which says that energy becomes increasingly "less available". In particular, the second law of thermodynamics concerns matter and energy; time itself is neither matter nor energy, so the second law of thermodynamics cannot tell us if time itself will come to an end or not. When pressed, therefore, (a) does not get us very far with respect to (i), (ii) and (iii) above.

At first, those who take the tack of saying (b) seem to be quite sophisticated because of the introduction of the notion of probability. Do not be deceived by this. We could ask them where the probability measure comes from, or how it is to be set up. If someone claims that one event is more probable than another, then that person has some measure that assigns a greater number to the possibility of that event than the other one. The number has to come from somewhere. We have to be comparing two things (events) and we need some unit of measure to come up with the numbers; and to compare their respective probabilities there has to be some plausible ground of comparison between the two which is some absolute, or fixed, frame of reference.

In the case above we say that time is probably potentially infinite, and presumably this means that it is less probable that time is finite. Is this a scientific claim or a conceptual claim? It cannot be a scientific claim, except in the rather shaky sense of there being more theories that postulate that time is potentially infinite. We are then counting theories. It is not obvious how to tell one theory from another, and it is not clear at all, given the past history of science, that the *number* of our present theories siding on one side, with respect to the infinity of time, is representative of reality. If we are not counting theories, then we are counting some sort of probability within a theory.

The problem here is that it is not obvious, mathematically, how to measure probability of time ending or not. There is no absolute background against which to measure the probability. So the term "probability", in the statistical or mathematical sense, is not appropriate here. At best, then, ascribing probability is just a measure of confidence, which is not quantifiable. If it really is not quantifiable, then our confidence is ungrounded.

A quick, but disingenuous, way of dealing with (c) is to point out that it is a tautology of the form "P or not P", where P can be replaced by any proposition or declarative sentence. Tautologies are always true, but they are also uninformative. More charitably, we could ask of (c) what "might" means, because "might" is often oblique for "has a probability measure". In this case, we return to the arguments over (b). On the other hand, if "might" is really to point to a sort of agnosticism, then it is possible for time not to come to an end, so it is possible for time to be infinite (actually!). So then we ask how we are to understand the possibility of actually infinite time. In doing so, we have uncovered a commitment to the notion of actual infinity at least as a possibility. So again, the concept of potential infinity is compatible with a conception of actual infinity. Again, this is something Aristotle would have rejected.

Statement (d) is an infuriating argument. It is not always legitimate, and we are entitled to ask where the "should" comes from. Is this normative or prescriptive? Is this a "should" of caution? Or is it a "should" of trying to cover up for the fact that the person using tactic (d) has nothing more to say? Disappointingly, this is often the case. Furthermore, the advocate of (d) can seldom defend the prescriptive or normative mode.

If we look closely, we see that these positions either beg the question, in the sense that the argument for the position is circular, or they rely on a conception of actual infinity. Thus we had better take a look at the notion of actual infinity a little more closely. To do so we shall investigate the mathematical notions of ordinal and cardinal infinities.

So far, we have dismissed arguments in favour of potential infinity in order to motivate looking at the notion of actual infinity. In Chapter 5, we shall return to more serious philosophical arguments in support of potential infinity as the only coherent notion of infinity. Henceforth, we shall refer to supporters of this viewpoint as "constructivists".[9] The arguments that constructivists give in favour of discarding actual infinity from mathematics revolve around two ideas. One is that mathematics is there to be applied to situations from outside mathematics, such as physics. There are only a finite number of objects in the universe, therefore, our mathematics should only deal with the finite. Call this the "ontological argument". The other motivation is more epistemic (having to do with knowledge). This idea is that there is no point in discussing infinite sets since we cannot know what happens at infinity or beyond infinity. More strongly, it is irrational to think that a person could be entertaining thoughts about infinity, since we are essentially finite beings, and have no access to such things. We shall revisit these arguments in Chapter 5.

4. The ordinal notion of infinity

The word "ordinal" in "ordinal notions of infinity" refers to the order of objects. A very intuitive example is that of people forming a queue. We label them as first in the queue, second in the queue, third in the queue and so on. The natural numbers, that is the numbers beginning with 1, followed by 2, then 3, then 4 and so on, are objects. They can be arranged in a *natural order* by using the same numbers as labels for "first", "second", "third" and so on: the natural number 1 is first in the order of natural numbers; the number 2 is second. We can discuss the ordinal numbers as a set of mathematical objects in their own right. The difference between the natural numbers and the ordinal numbers can be confusing; remember simply that the natural numbers are conceptually prior to the ordinals. The natural numbers are quite primitive, and they are what we first learn about. We then transpose a (quite sophisticated) theory of ordering on them. For convenience we use the natural numbers in their *natural order* to give order to any set of objects we can order. We use the ordinals (exact copies of the natural numbers) to order objects such as the natural numbers. So the order is one level of abstraction up from the natural numbers: we impose an order on objects.

The finite ordinals can be gathered into a set in their own right. The set is referred to as "the set of ordinal numbers". This makes for a certain amount of ambiguity in referring to the ordinals as labels, as a series of numbers or as a set of numbers. For our purposes, it is more important to think of the ordinal numbers as a well-ordered series of labels.[10] The natural numbers are ordered by the "less than" relationship, often symbolized "<". When we say that the natural numbers are ordered by "<", what we mean is that given any two distinct numbers one is strictly greater than the other.

There are all sorts of ordering relations to which we might want to give mathematical expression. For example, we might want to capture mathematically the idea of temporal order. For example, we might ask: did Achilles cross the finish line before or after the tortoise? We can order people in terms of height, so Paul might be taller than Bert. We can order physical objects in terms of volume: "this pot is bigger than that pot" usually means that the first can hold more liquid (in terms of volume). Provided we have a comparative measure,[11] we can label the order of things. More simply, provided we have an ordering relation we can order things.

Let us return to the natural numbers. These can "order themselves", in the sense of clearly revealing their own natural order. Recall that the natural numbers are ordered by the relation "<". But the natural numbers are infinite. At "the end" of the natural numbers we have a new number called "omega", the last letter of the Greek alphabet, the symbol for which is "ω". Now ω is an unusual number: it is the first infinite ordinal. It is an ordinal because it is strictly greater than any finite ordinal number, so it follows in the "strictly greater than" series. However, it lacks an immediate predecessor: a number that comes directly before it. The one finite ordinal that shares this feature is 1. In the series of natural numbers 1 has no immediate predecessor;[12] ω has no immediate predecessor and is the successor of all the finite ordinals. That is, there is no number ω − 1. If there were, then we could work our way back to the finite numbers, and then ω would be finite. Because ω has no immediate predecessor, we call it a "limit ordinal".

Although ω has no immediate predecessor, it does have an immediate successor: ω + 1. This is because, by definition, all the ordinal numbers have an immediate successor. Suddenly we have two infinite ordinals. It gets better. Since we have ω + 1, we also have ω + 2, ω + 3 and so on. What happens at "the end" of this part of the series of ordinals? We can add ω to ω, which is the same as 2 × ω. This is another limit ordinal. It has no immediate predecessor, but it does have an immediate successor: (2 × ω) + 1. The next limit ordinal is 3 × ω. In addition to finding limit ordinals in this way, we can also take powers of ω, for example, ω × ω, written ω^2. The limit ordinal of the series of ω raised to successive powers of ω – that is, ω, ω^ω, ω^{ω^ω}, … – is given a new letter ε (the Greek letter "epsilon"). We can continue combining ε with finite numbers, ω and ε itself by addition, multiplication and exponentiation. This gives us an idea of how mathematicians extend the ordinal numbers into infinity, and beyond the least infinite ordinal, ω. What does the order of all these extensions look like?

1, 2, 3, …, ω, ω + 1, ω + 2, …, 2ω, 2ω + 1, 2ω + 2, …, 3ω, 3ω + 1, 3ω + 2, …, ω^2, ω^2 + 1, ω^2 + 2, …, ω^3, ω^3 + 1, ω^3 + 2, … ω^ω, ω^ω + 1, ω^ω + 2, …, (= ε, so we can continue), ε + 1, ε + 2, ε + 3, …

Note that "," indicates that the next number is the immediate successor, and "…" indicates that at least one limit ordinal follows. Not only do we have an infinite ordinal, we have an infinite number of infinite ordinals.

5. The cardinal notion of infinity

To spark the imagination, and introduce the concept of infinite cardinals, it is customary to relate some version of the following story, which is based on the mathematician David Hilbert's (1862–1943) hotel paradox. Consider a hotel with an infinite number of rooms. A large conference on mathematics is to take place, and all the delegates are to be accommodated in the hotel. They start to arrive the day before the conference and are allocated rooms in order: room 1, room 2, room 3 and so on. On the first day of the conference, more delegates arrive, an infinite number of them, and the hotel is able to accommodate them. But then there is a problem. A tourist now arrives in town, and there is only the one hotel, with an infinite number of rooms, currently occupied by an infinite number of conference delegates. The tourist asks for a room for the night. The receptionist could ask her to take the room at the end, but this would involve walking an infinitely long way. Instead, the receptionist finds another solution, asking everyone in the occupied rooms to move to the next-numbered room. This frees up the first room, which is where the tourist is accommodated. Had an infinite number of tourists arrived, the receptionist could have asked all the conference delegates to move to the even-numbered rooms found by doubling their original room numbers, thus freeing up an infinite number of odd-numbered rooms for the tourists. There would always be room for more in this hotel! Infinite cardinals can "absorb" finite and some infinite cardinal numbers without changing. How can this be?

The cardinal numbers answer the question "How many?"; the order of presentation of the objects being counted is immaterial. For example, two sets of three objects have the same cardinality: the cardinality of each set is three. It does not matter if the objects in one set are much larger than those in the other set; we just count the members of the sets. A set is indicated by curly brackets, and the members of the set are written inside the brackets, separated by commas. Let A be the set containing the numbers 6, 95 and 62. Then A = {6, 95, 62}. Similarly, let B = {567, 2, 1346}. Both A and B have cardinality 3.

Definition: The cardinality of a set is the number of members of the set.

By the "cardinality" of a set we mean the "size" of a set. Two sets are "of the same size" if they have the same cardinality: A and B are of the same size.

Now that we have a notion of cardinality, and one of "sameness of size", we can consider two different infinite sets of numbers. One is the set of *all* finite natural numbers, the other the set of all even numbers. Do both of these sets have the same number of members? The obvious first answer is that the set of even numbers has fewer members than that of all the natural numbers. It is missing half the numbers, so it must have cardinality half of infinity. But "half of infinity" is a peculiar answer. Maybe half of infinity is also infinity: think of Hilbert's famous hotel. We need to think about how to compare the cardinalities of infinite sets. To do this we need some more definitions.

Definition: A subset of a set A is a set containing only members of A. The empty set is a subset of every set. Note that it is not a *member* of every set.

A subset may include all the members of the original set, or it may leave some out. The empty set is the set with no members (cardinality 0). Its being a subset of every set follows trivially from the definition of subset; the subset of any set has to include the set of nothing at all. We denote the empty set with the symbol "∅".

Definition: A proper subset of a set A is a subset that does not contain all the members of A.

So a proper subset of a set is a subset that is missing at least one member of the original set. For example, consider the set of finite natural numbers, {1, 2, 3, 4, 5, ...}. A proper subset of this set is that of all the finite natural numbers beginning with 5: {5, 6, 7, 8, ...}. This is a subset because it only includes members of the original set and it is a proper subset because it is missing at least one of the original members (in fact it is missing four, the numbers 1, 2, 3 and 4).

If we can match two sets up so that we can pair off each and every member of the first set with one, and only one, member of the second set, then we have placed the two sets in one-to-one correspondence.

Definition: Two sets, A and B can be placed in one-to-one correspondence just in case every member of A can be matched up with a unique member of B and *vice versa*.

When we can do this, we say that the two sets are *of the same size*.

Definition: Two sets are of the same size if and only if they can be placed in one-to-one correspondence.

Recall that we asked whether the set of finite natural numbers was the same size as the set of even numbers. Note that we are just thinking of these as sets, not as ordered series. Call the set of natural numbers $\mathbb{N}$ and the set of even numbers E. Now, E is a proper subset of $\mathbb{N}$; it is missing all the odd numbers. However, E can be placed in one-to-one correspondence with $\mathbb{N}$. We can pair up 1 from $\mathbb{N}$ with 2 from E, 2 from $\mathbb{N}$ with 4 from E, 3 from $\mathbb{N}$ with 6 from E. We can carry on this pairing indefinitely because both sets are infinite:

```
ℕ    1    2    3    4    ...
     ↕    ↕    ↕    ↕
E    2    4    6    8    ...
```

The set of natural numbers, ℕ, can be placed in one-to-one correspondence with one of its proper subsets, E. Therefore, ℕ and E are of the same size.[13]

We now have enough concepts to introduce a historically important definition: Richard Dedekind's (1831–1916) definition of an infinite set.[14]

> *Definition:* A set is infinite if and only if it can be placed in one-to-one correspondence with one of its proper subsets.

So the set ℕ is infinite. This will not work with any finite set (try some examples), but will, of course, work with any infinite set.

Are there other sets that are infinite by Dedekind's definition? The set of even numbers, E, is also infinite, because we can find a proper subset of E that we can place in one-to-one correspondence with it. Consider, for example, the set C of all even finite numbers except 2. C is a subset of E because it only contains members of E; it is a proper subset of E because it is missing 2. And E can be placed in one-to-one correspondence with C in the following way: match 2 of E with 4 of C; match 4 of E with 6 of C; match 6 of E with 8 of C and so on.

```
E    2    4    6    8    ...
     ↕    ↕    ↕    ↕
C    4    6    8    10   ...
```

The way in which we find the matching is irrelevant; we just have to show that there is one. Dedekind's definition of infinity distinguishes finite from infinite cardinal numbers.

We can now move on to ask: are all infinite sets of the same size? To answer this we shall have to address more sophisticated notions among the infinite cardinal numbers. We shall start with sets that we intuitively think are bigger. Consider the set of integers, ℤ. This is all the negative natural numbers and 0 together with the positive natural numbers: … −3, −2, −1, 0, 1, 2, 3, …. The (positive) natural numbers form a proper subset of the integers. Are the two sets ℕ and ℤ of the same size? That is, can the two sets be placed in one-to-one correspondence with each other? We might think not, because the integers go on infinitely in two directions, not just one, which suggests that there are twice as many integers as there are natural numbers. But think again.

Remember that when we are interested in the cardinality of a set, we are only interested in answers to the question: how many? As such, we can disregard the usual order of the numbers. We could match 0 from the set of integers to 1 of the set of natural numbers, 1 from the set of integers to 2 from the set of natural numbers, −1 from the set of integers to 3 from the set of natural numbers, 2 from the set of integers to 4 from the set of natural numbers, −2 from the set of integers to 5 from the set of natural numbers, and so on.

Natural numbers	1	2	3	4	5	6	7	...
	↕	↕	↕	↕	↕	↕	↕	
Integers	0	1	−1	2	−2	3	−3	...

Since we have placed the set of integers in one-to-one correspondence with the natural numbers, we say that the set of integers is of the same size (or has the same cardinality) as the set of natural numbers.

What about the rational numbers? Rational numbers are all those that can be represented in the form m/n (i.e. as fractions) where m and n are natural numbers different from 0.[15] Between any two rational numbers there is a third rational number, and it follows that there is an infinite number of rational numbers between any two rational numbers. For example, between ¼ (= 6/24) and ⅓ (= 8/24) there is 7/24, and between ¼ (= 12/48) and 7/24 (= 14/48) there is 13/48, and so on. To describe this, we say that the rational numbers are "everywhere dense".

We might think that there must be more rational numbers than natural numbers. In fact, since between any two rational numbers there is an infinite number of rational numbers, we could consider that we have infinity in three dimensions: positive numbers, negative numbers and the "depth" of an infinite number of rational numbers between any two rational numbers. But it is possible to place the set of rational numbers in one-to-one correspondence with the natural numbers so (from the definition above) the sets of rational numbers and natural numbers are the same size.

To see this consider Figure 10, which gives a tabular representation of the rational numbers. We first have to agree that, if completed, this table will capture all the possible rational numbers. Of course, completing the table is a superhuman task, but that does not matter. We can see that all the numbers are represented: nothing is missed out. Note that we miss out 0 on the vertical vertex because anything divided by 0 is "undefined". Also note that there are repetitions: ½ is the same as 2/4, is the same as 3/6 and so on. We can eliminate the repetitions as we go along by converting each rational into its "lowest form" and then erasing exact copies. How are we going to show that this table can be put into one-to-one correspondence with the natural numbers?

For added simplicity, ignore the fact that there are repetitions. We can draw a continuous line through all the entries on the table by starting in the middle and spiralling outwards (Fig. 11). We should agree that all the numbers

	...	-3	-2	-1	0	1	2	3	...
		⋮	⋮	⋮	⋮	⋮	⋮	⋮	
3	...	-3/3	-2/3	-1/3	0/3	1/3	2/3	3/3	...
2	...	-3/2	-2/2	-1/2	0/2	1/2	2/2	3/2	...
1	...	-3/1	-2/1	-1/1	0/1	1/1	2/1	3/1	...
-1	...	-3/-1	-2/-1	-1/-1	0/-1	1/-1	2/-1	3/-1	...
-2	...	-3/-2	-2/-2	-1/-2	0/-2	1/-2	2/-2	3/-2	...
-3	...	-3/-3	-2/-3	-1/-3	0/-3	1/-3	2/-3	3/-3	...
		⋮	⋮	⋮	⋮	⋮	⋮	⋮	

Figure 10

are collected using this line. We can now list the numbers according to their order on that line – $0/_{-1}$, $1/_{-1}$, $1/_1$, $0/_1$, $-1/_1$, $-1/_{-1}$, $-1/_{-2}$, $0/_{-2}$, ... – and then put all the numbers into their lowest form, eliminating repetitions. We then have: 0, −1, 1, ½, −½ ... . We can now place these in one-to-one correspondence with the set of natural numbers:

Natural numbers	1	2	3	4	5	6	7	...
	↕	↕	↕	↕	↕	↕	↕	
Rational numbers	0	−1	1	½	−½	−2	2	...

You might think this is cheating somehow, but recall from the definitions above that cardinality is a measure of "how many", regardless of the order of the numbers. To show that there is a one-to-one correspondence, we only have to give one way of setting up the correspondence. It is sufficient that we agree that all the rational numbers will eventually be enumerated using this method. Clearly, and trivially, there are alternative methods.

How do we denote the cardinalities of infinite sets? Finite sets have natural-number cardinalities: a set with three members has cardinality 3; one with eighty-nine members has cardinality 89 and so on. We have names for infinite cardinalities too. The first infinite cardinal is $\aleph_0$. The symbol $\aleph$ (pronounced "alef", and written "aleph" in English) is the capital version of the first letter in the Hebrew alphabet. The "0" is the number zero. To talk of $\aleph_0$ we can say "aleph-zero", "aleph-null" or "aleph-nought".

Having demonstrated that the set of rational numbers can be put into one-to-one correspondence with the set of natural numbers, we might now think that all infinite sets have the same cardinality. The cardinality of the set

	...	−3	−2	−1	0	1	2	3	...
	...	⋮	⋮	⋮	⋮	⋮	⋮	⋮	
3	...	-3/3	-2/3	-1/3	0/3	1/3	2/3	3/3	...
2	...	-3/2	-2/2	-1/2	0/2	1/2	2/2	3/2	...
1	...	-3/1	-2/1	-1/1	0/1	1/1	2/1	3/1	...
−1	...	-3/-1	-2/-1	-1/-1	0/-1	1/-1	2/-1	3/-1	...
−2	...	-3/-2	-2/-2	-1/-2	0/-2	1/-2	2/-2	3/-2	...
−3	...	-3/-3	-2/-3	-1/-3	0/-3	1/-3	2/-3	3/-3	...
		⋮	⋮	⋮	⋮	⋮	⋮	⋮	

Figure 11

of natural numbers is $\aleph_0$, and this is also the cardinality of the set of even numbers, the set of integers and the set of rational numbers. But there is at least one set of numbers that has a greater cardinality than $\aleph_0$.

The set of real numbers consists in the set of rational numbers together with the set of irrational numbers. A number is irrational just in case it cannot be represented as a fraction: in the form $^m/_n$. An irrational number has an infinite non-repeating decimal expansion (i.e. the numbers after the decimal do not form a pattern that is exactly repeated). For example, 0.12112111211112 ... forms a pattern but not a pattern that is exactly repeated, so it is an irrational number. In contrast, 0.333333... is a rational number (it can be written ⅓) as is 8.345345345345... . The exactly repeating pattern part of this decimal expansion is "345". We call this the "period" of the decimal expansion. Any number with a period in its decimal expansion can be represented as a fraction. The number 1.18181818... , for example, can be represented as the fraction ¹³⁄₁₁. Famous examples of *irrational* numbers are π (the Greek letter "pi"), and e.[16]

We say that the set of real numbers, consisting in the set of rational numbers *and* the set of irrational numbers, represents the "continuum". We think of the line formed by the real numbers as smooth, having no gaps; the set of rational numbers has gaps, since it is missing all the irrational numbers.

The first thorough mathematical treatment of infinite cardinal numbers was developed by Georg Cantor (1845–1918) in the 1880s, and he proved that the set of real numbers has a cardinality greater than $\aleph_0$. Part of the proof technique was new. It is called "diagonalization", and has been replicated in many proofs in mathematics since. We shall work through the proof here. It is not difficult to follow, although it is tempting to think at the end that there has been some sleight of hand.[17]

The overall structure of the proof is a *reductio ad absurdum* argument, in which we (i) suppose that the set of real numbers is the same size as the set of natural numbers and then (ii) show that this leads to a contradiction. We then (iii) conclude that the two sets must be of different sizes. Since it is obvious that, of the two, the set of natural numbers is smaller than that of the real numbers, because the natural numbers form a proper subset of the real numbers, we (iv) conclude that the set of real numbers must have a greater cardinality.

Step (i): Suppose (for the sake of argument) that the set of real numbers is the same size as the set of natural numbers.

Step (ii): Under the assumption of (i), there should be some way of listing these in some order, so that they can be placed in one-to-one correspondence with the natural numbers. Let us concentrate for now on just the real numbers between 0 and 1. Each of these has an infinite decimal expansion. Some of these will be trivial, for example, 0.500000... . Some will be repeating; some not. We could list them in supposed order in the following format, where the subscript numerals refer to the digit in the decimal, and the superscript numerals refer to the ordering of the real numbers:[18]

1   $0.a_1^1 a_2^1 a_3^1 a_4^1 a_5^1...$
2   $0.a_1^2 a_2^2 a_3^2 a_4^2 a_5^2...$
3   $0.a_1^3 a_2^3 a_3^3 a_4^3 a_5^3...$
...   ...

(For example, if the third number in our list is 0.236835 then a_1^3 refers to 2, a_2^3 refers to 3, a_3^3 refers to 6 and so on.) Now consider a number made up from the digits along the diagonal, $a_1^1 a_2^2 a_3^3$.... Call this the "diagonal number". Since we are listing *all* the real numbers, it should turn up in the list at some point. Now let us modify the diagonal number. For each digit in the diagonal number, add 1 to it, unless it is 9; if it is 9, then turn it into the digit 1. Call this new number "Cantor's diagonal number". Cantor's diagonal number will not turn up on the list above: it is different from the first number on the list in at least the first digit; it is different from the second number at least at the second digit; it is different from the third number on the list at least at the third digit and so on. It does not appear in the original list at all.

Recall that we had to be able to list *all* the real numbers between 0 and 1 in order to place them in one-to-one correspondence with the natural numbers, which we know we can list. We have generated a contradiction to our original supposition that we could list the real numbers in this way.

Step (iii): Two sets of numbers have the same size (i.e. same cardinality) if and only if they can be placed in one-to-one correspondence with each other. We conclude that because the set of real numbers between 0 and 1 cannot be placed in one-to-one correspondence with the natural numbers, the set of real numbers between 0 and 1 must be of a different size than the set of natural numbers. *Mutatis mutandis* for the whole set of real numbers.

Step (iv): The natural numbers form a proper subset of the set of real numbers, so we can conclude that the set of real numbers is larger than the set of natural numbers. In fact, even the set of real numbers *between 0 and 1* is larger than the set of natural numbers.

What are the implications of this proof? We call $\aleph_1$ the next cardinality up from $\aleph_0$. But we do not know if the continuum (the whole set of real numbers) is exactly one size up from the size of the natural numbers; whether the continuum is of size $\aleph_1$ or bigger. This is called "the continuum problem".[19] It is provably independent of Zermelo–Fraenkel set theory, which is the current standard accepted set theory. That is, we can have a consistent set theory where the continuum has cardinality $\aleph_1$, and we can have a consistent set theory where the continuum is represented by a higher cardinality. The problem of deciding whether the continuum is the next cardinal after $\aleph_0$, and so should be represented by $\aleph_1$ is a deep and difficult one. There seems to be no good mathematical way to decide the continuum problem because the notion of the continuum (or the "seamless" or "gapless" line) is an informal notion. The continuum, like the notion of "natural number" comes before formal representation in set theory. When we develop a formal set theory, we try to capture the informal notion as best we can. There are competing definitions of "real number", and therefore of the collection of all real numbers in the set called the continuum. Some definitions have fallen into disuse, but they are consistent with some standard set theories. We need not go into the technicalities here. What is important is that the formal representation of the continuum is provably independent of the theory of infinite cardinals, and the theory of how we get from one infinite cardinal to the next. In standard set theory we get to the next cardinality from one infinite cardinality by taking the powerset of the infinite cardinality in question.

Definition: The powerset of a set is a set made up of all the subsets of a set.

For example, the powerset of the finite set {9, 654, 24} is {∅, {9}, {654}, {24}, {9, 654}, {9, 24}, {654, 24}, {9, 654, 24}}. The operation of taking the powerset of an infinite set gets us from one infinite cardinality to the next. That is, $\aleph_1$

is equal to the powerset of $\aleph_0$, which is equal to $2^{\aleph_0}$. This powerset operation is independent of our definition of a real number, and thus of the continuum (as the set of all the real numbers). The present consensus over the continuum problem is that the set of real numbers is either of the size $\aleph_1$ or it is greater than $\aleph_1$. Which it is can be decided by adding axioms to the basic axioms of Zermelo–Fraenkel set theory, but there are several possible axioms to add, and both results can be obtained.

The idea of the continuum is important not only in mathematics, but also in physics. Physicists and metaphysicians usually think of space and time as continuous in the sense of seamless or gapless. Only the set of real numbers can mathematically represent this. But the continuum problem shows us that we cannot really know where the continuum fits in our theory of the cardinal numbers.

6. Summary

We now know enough about infinity to engage the philosophical disputes in the philosophy of mathematics. The important points to retain from this chapter are:

- the distinction between potential and actual infinity;
- the distinction between ordinal numbers and cardinal numbers;
- the idea that, in current mathematics, it is widely accepted that there are different sizes of infinity, and that there are an infinite number of infinite ordinals;
- an appreciation that there are several pre-mathematical conceptions of infinity, and that there are several mathematical conceptions too.

Chapter 2
Mathematical Platonism and realism

1. Introduction

Together with Chapter 1, this is a lynchpin chapter in this book. Most positions in the philosophy of mathematics can be cast as reactions to Platonism. Some of them, such as logicism or some forms of structuralism, are modifications of Platonism; some, such as constructivism, are very strong reactions to it. It is important to study this chapter well, and have a solid grasp of realism by the end of it.

Section 2 will discuss Plato, since he is the originator of the realist position in the philosophy of mathematics. In §3 we then turn to realism more generally, as the modern incarnation of Platonism. We shall then look at two important modern defenders of realism: Kurt Gödel (§4) and Penelope Maddy (§5). We then discuss some very important problems with the views defended by both Gödel and Maddy, in different ways and to different degrees. Section 6 on the problems with set-theoretic realism motivates the rest of the book.

2. Historical origins

Unsurprisingly, mathematical Platonism originated with Plato (*c.* 427–*c.* 347 BCE). Plato was interested in what mathematical truth consists in. He was exposed to geometry and arithmetic, so he was interested in what secures geometrical and arithmetical truths.

Plato observed that we grasp geometrical theorems in a way that is quite different from the way in which we grasp empirical truths, which are truths we arrive at using our senses. To learn mathematics we need neither have much sense experience nor be taught particular formulas, in the sense of memorizing them. It is enough for us to learn a few general principles, and we can piece together what we need to solve particular problems. In other words we seem to be able to reason *a priori* about geometry and arithmetic.

In one of Plato's famous dialogues, the *Meno*, Socrates[1] proposes to his interlocutor, Meno, that he run an experiment to illustrate his thought about how we come to know about geometrical truths. The experiment is to ask someone who has little training in geometry to develop a theorem of geometry himself with only a little rational guidance. The chosen person is a slave in Meno's household. The slave has had little schooling in geometry, but he can read, write and knows what a triangle is, for example. Socrates does not tell the boy how to work out the problem he poses, but he does indicate to the boy when it is that he has made a mistake. The slave develops Pythagoras's theorem: that the square of the hypotenuse of a right-angled triangle is equal to the sum of the squares on the other two sides. It is quite an impressive feat, and quite believable. With this experiment Socrates demonstrates that we need no special prior knowledge in order to develop theorems in geometry. Moreover, we do not need to be told what they are; we can develop them ourselves provided we are aware of when we are reasoning poorly. Moreover, it seems that any rational person can help us see that we are reasoning poorly. Good reasoning is universally recognized.

The well-informed reader might object at this point. For the modern reader it seems fantastic to claim that the geometry we develop under rational guidance will turn out to be *the truths of geometry*, since geometry includes the study of many different geometrical systems. We need to be careful. Today, we know that there are alternative geometries that were developed in the late-nineteenth century and later. It would be presumptuous to say that, in so far as they differ from Euclidean geometry, they are false. It is more diplomatic to say that there are different geometries. Each contains truths, that is, axioms and theorems of the system. So why was Euclidean geometry the only geometry for so long? The mathematical community was ready to accept alternatives only after the proof about the independence of Euclid's fifth "parallel" postulate was proved.

The fifth postulate says that if there is a straight line (infinitely extended in both directions), and a point not on the line, then there is only one straight line (infinitely extended in both directions) that runs through the point and will never intersect with the first line (Fig. 12). This new line is parallel to the first. In 1868 Eugenio Beltrami (1835–1900) proved that this postulate was independent of the other axioms of Euclidean geometry,[2] that is, that it was possible to have a consistent system using the first four axioms and denying the parallel postulate. Nikolai Ivanovich Lobachevsky (1792–1856) and, independently, Jànos Bolyai (1802–1860) had already developed hyperbolic geometry in 1829,

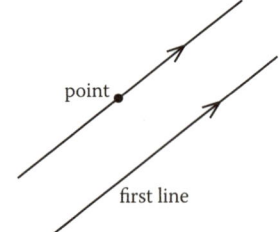

Figure 12 The two lines are parallel. They will never intersect.

but they had not proved explicitly that this was a non-Euclidean geometry (one that does not use the parallel postulate). These discoveries suddenly opened the door to the possibility of other axioms of geometry, each independent of the four original axioms of Euclid, the addition of which made for new systems of geometry. Thus, we do not now study just one theory when we study geometry. We might study any number of geometrical systems, all emanating from Euclidean geometry.[3] For the sake of following Plato's development of Platonism in mathematics, we shall temporarily ignore the recent developments in geometry, and pretend that the only geometry is Euclidean geometry. This is perfectly legitimate. The first four postulates are common to all theories of geometry. (Otherwise we are studying topology, or some other branch of mathematics.) Moreover, Euclidean geometry is intuitive and easy to picture.

The interesting question is how is this grasping of (Euclidean) geometrical truths possible? For Plato, and Socrates, the very possibility of being able to reason *a priori* about geometry depends on our being able to reason about something. In this case the "something" is abstract, not concrete or physical. In particular, we are not reasoning about particular triangles drawn in the sand (or, more recently, on paper), but quite generally about triangles. Triangles, and other geometrical figures, do not seem to depend on particular drawings of triangles; in fact, our drawings of triangles depend on an idea of a perfect triangle to which we aspire. The idea of the triangle is not subjective, in the sense of varying from one person to the next; rather, it seems to be a universal idea that exists independently of our choice or ability to conceive it or draw it.[4] But what do we mean by "exists" in the last sentence? Socrates and Plato developed a theory, which we call "Platonism", that there exists a realm of perfect objects quite independent of human beings. The objects in "Plato's heaven" are perfect, and everything on earth is a pale imitation of them. The objects are called "Forms" or "Ideals", depending on the translation.

The relation between the Form and its pale physical imitations on earth is somewhat analogous with attempts at realistic art. A realistic sculpture of a person is never a perfect representation of the person. Inevitably, there are some small mistakes or differences. Similarly, when we try to draw a triangle, it is a pale imitation of the perfect triangle in Plato's heaven. When we reason about geometrical figures, we reason about perfect ideal objects, not about drawings. This is what guarantees that we can be quite general when we so reason.

Let us return to the issue of our grasping geometrical truths. Platonist philosophy says that as human beings we have insight into this realm of perfect forms, and this is how we can aspire to draw more and more accurate triangles. Moreover, this is how we can determine that one illustration of a triangle is better than another. We can make these comparative judgements because we compare particular drawings to the perfect triangle.[5] The perfect triangle is something our mind has access to, not our bodies; that is, we do not sense perfect triangles using our five senses. Instead, we use our reasoning, our intuition: in short, our mind.

Platonist philosophy is plausible. Many mathematicians tend towards Platonism. They have a clear sense of grasping mathematical truths, of understanding something that is true independently of us, is eternally true and is very present to the mind. Furthermore, if taken at face value, the language used by mathematicians presupposes a sort of Platonist conception.[6] Mathematics texts talk about discovering truths and finding the proof for a true theorem; the sentences suggest that there is a realm of truths out there, and we have the job of understanding them. When a mathematician talks about "concrete" examples, she is not referring to physical examples. Rather, she is referring to Ideal natural numbers, or Ideal triangles. This is all Platonist talk. The opposite would be constructivist talk, where one creates the truths, or constructs some mathematical object. For the constructivist, we do not discover mathematical objects ready-made; we create or construct them.

We shall start to use the lower-case "platonism" to indicate a philosophy that shares many features with Platonism, but is not Platonism in the sense of following Plato no matter what he might say. A Platonist is someone interested in what the great master Plato thought, and this same person will believe what Plato said, and take it as truth. A platonist is inclined towards Platonism, but is willing to modify it, and will not make special reference to Plato. A realist, is someone who shares his, or her, views with the platonist, but who will express the view in vocabulary more usually associated with realism than with Plato. The Platonist, platonist and realist have overlapping positions. For our purposes, we are more interested in platonism and realism than Platonism and interpretations of Plato. Bearing this in mind, let us now turn to the realist view, which is the modern incarnation of platonism.

3. Realism in general

From ancient Greece we now jump to the twentieth century. This is because the current philosophy of mathematics was developed largely as a result of the discoveries of the set-theoretic paradoxes at the beginning of the twentieth century. These caused a crisis in the foundations of mathematics, which led to the current philosophical reflection concerning mathematics. Let us illustrate with the Burali-Forti paradox, which involves infinite ordinals.[7]

The ordinals, which we encountered in Chapter 1 as the set of numbers used to "order" objects, can be described in another way. We can define an ordinal as the set of all preceding ordinals, very elegantly building them up from the empty set. The idea is this. We stipulate that ∅ is the first ordinal, corresponding to the word "first". The set of this is the next ordinal, corresponding to the word "second", represented {∅}. To form "third", we gather the two previous ordinals into a set: {∅, {∅}}. "Fourth" is formed by gathering the three previous ordinals into a set: {∅, {∅}, {∅, {∅}}}; and so on. We can carry out this procedure indefinitely, which is what we want, since there are an infinite number of ordinals. This way of defining the ordinals is elegant because we can always tell which of two ordinals is the greater since the smaller will be a member of the greater ordinal. We can also tell if two ordinals are identical: they have the same members. Furthermore, all the ordinals are built up from the very sparse empty set: ∅.

Now we encounter a problem. Since we are making ordinals by this "gathering" procedure, we can consider the set of all the ordinals: the culmination of this gathering is all the ordinals gathered together. Recall that we have made the ordinals by swallowing up all the previous ones. Now, is the set of all ordinals an ordinal? If we think it is, then it should be included in the set of all the ordinals, so we do not have a complete set. If we think it is not, then it should be because it is indistinguishable from other ordinals, and therefore should be included. Contradiction.

How does the realist react? There is a technical solution to the paradox. We distinguish betwen sets and classes. A class is a collection of sets. All sets are classes, but not all classes are sets. Classes that are not sets are called "proper classes". These are not constructed from, or derived from, the set axioms, which are quite conservative about what can be built up. This distinction between a set and a proper class has two immediate repercussions. One is that we are no longer allowed to refer to the "set of all ordinals": it is a "proper class of all ordinals". Apart from the linguistic difference, there is a very powerful technical difference. We have procedures for building up sets from other sets, but none of the building procedures will get us to a proper class. Proper classes are those things we discuss when we say consider "all" the so-and-sos when these cannot be reached by "normal" set-construction methods. We can think of it this way. The axioms of set theory allow us to build sets from previous sets, but the axioms of set theory do not tell us how to construct a proper class. They do, however, give us some properties of proper classes. The building process for sets is conservative. When we talk of proper classes we look at ideas we can form in the language. So the only limitation is formal grammar. "Proper class" talk is from the top down: "set" talk is from the ground up. We get into trouble, in the form of paradoxes, when we confuse the two approaches. The realist is at home with this, for this technical solution indicates that when we developed the set-theoretic paradoxes from what is now called "naïve" set theory, we were confused. We did not know enough. Now we know more, and have adjusted our set theory so as not to engender paradox, so we see more clearly. The class–set distinction allows the realist to breathe easy for now. We shall not discuss more about the set-theoretic paradoxes here, but shall see more later, since reactions to these motivated many of the current philosophical positions in the philosophy of mathematics.[8] Instead, let us develop our understanding of realism as a philosophical position.

To begin with, in the way in which the terms are used here, "platonist" and "realist" are interchangeable. We shall stick to "realist", but the reader should be aware that sometimes what we are referring to as realism is called "platonism" in the literature. There are many versions of realism. Let us look at the points of divergence between realist positions. One example of a way in which two realists might diverge in their positions is over the notion of being a realist in ontology or a realist in truth-value.[9] What realists all have in common is the idea that the truths of mathematics are not of our making. The contention is over the objects. The realist in ontology thinks that there are a number of independent objects. This is what makes our judgements about them objective. The realist in truth-value believes that the truths of mathematics are independent of us, but not necessarily that what makes them independent is a realm of objects. The realist in truth-value might be agnostic as to what the truth-makers of sentences are, or might think that something other than "objects", in any strong sense of "object", makes the truths of mathematics true independent of us. Intuitively, this makes for an uncomfortable position. For, there is nothing to make the sentences true, and yet they are true independent of us. Many philosophers do not like to put independence of truth together with agnosticism about ontology, so they resist the position. On the other hand, it does have some merit. The realist in truth-value can be indifferent to there being any objects corresponding to the truths of mathematics. This saves the philosopher having to explain anything about abstract objects. The reasons for doing this are, again, intuitive, and build on the notion that the canonical example of an object is a physical object. The things that exist are physical objects. The rest are either nonexistent, or we are agnostic about what their status is. The realist in truth-value argues that it is difficult to imagine that timeless ethereal objects such as numbers are objects in just the same way that tables and chairs are objects. Most realist positions are both realist in ontology and in truth-value, since a realism in ontology explains, or justifies, realism in truth-values. This is because it is the independent objects that make mathematical truths true.

Another point of divergence between realist positions concerns what it is that we are realists about. This point occurs between different realists about ontology. That is, some realists are realists about sets; others are realist about numbers; others are realist about the geometrical primitives: points, lines and planes. To argue for one mathematical ontology over another, one has to argue that the theory that has these objects as primitive is a "founding discipline" within mathematics. A "founding discipline" is such if we can reduce other parts of mathematics to it. The founding discipline is sometimes referred to as the "reducing discipline". To show a reduction of one discipline, or area of mathematics, to another, we have to show that we can translate from the reduced discipline to the reducing discipline. We also have to show that we can recapture the truths of the reduced discipline in the reducing discipline. Successful competing founding disciplines in mathematics include: the various set theories, type theory, category theory, model theory and topology. The very fact that there are competing founding disciplines provokes further discussion as to the merits and faults of each.

A third point of divergence between realists (of the ontological or the truth-value stripe) concerns epistemology. Different realist theories will distinguish themselves from each other by appeal to different theories about how we can know the truths of mathematics. Some will appeal to intuition (see Plato); others argue that we perceive mathematical truths (see Maddy); still others argues that our knowledge is analytic and *a priori*, and therefore does not depend on intuition (in a Kantian sense of the term "intuition") (see Frege). We shall be discussing and refining the various readings of "intuition" throughout the text. For now, it is enough to know that there are several.

Equipped with these three points of divergence within realism, let us take a broader view and look at some of the arguments for and against realism. To do so, we should refine our vocabulary. In giving a philosophical theory, it is often useful to distinguish between two parts of the theory: the ontological part and the epistemological part. The ontological part concerns what there is according to our theory: what objects exist according to our theory. The epistemological part concerns how it is that we know about the objects postulated by our theory.

The ontological part of Plato's theory of mathematics concerns Plato's heaven of perfect Forms. The epistemological part concerns this notion of reasoning *a priori* (without direct appeal to the senses) about mathematical objects. The two parts of Plato's philosophy of mathematics compliment each other rather well, and are adopted by the modern realist (in ontology). One part of the Platonist theory explains the other: we can explain the epistemology in terms of the ontology, and the ontology in terms of the epistemology. More explicitly, to account for how it is that we know (i.e. epistemology) independent truths without appealing to sense experience, we think that there must be something that makes the truths true. That something is a realm of objects (i.e. ontology) that exists independently of us, and that is filled with perfect objects. The other way round is that if we accept that there is a realm of independent mathematical objects, then we have access to these in a pure and perfect manner: we do not need to appeal to our experience or our senses, since everything that we sense is imperfect, because it is down here on earth and subject to distortion. We have direct insight to perfect mathematical objects. Sometimes, in modern literature, this insight is called "intuition". Reason helps guide and correct our intuitions.

According to the modern realist (in ontology), we reason about these objects, and we intuit them. The notion of intuiting mathematics will crop up quite often as we examine the alternative philosophies of mathematics. The reader should be careful whenever this word is seen in the context of the philosophy of mathematics, and make sure that she possesses a good understanding of the definition of the word in that context.

So what do we mean by intuition, at least at this stage in our enquiry? According to the realist, we seem to have access to a realm of perfect objects, or Forms. We "see" them in our mind's eye. They are perceivable by us. Mathematical objects sit before our minds as vividly as do physical objects such as tables and chairs, once we have learned to see them. It is one of the privileges of being human that we can intuit mathematical objects.[10] We use them to judge whether or not a mathematical description is accurate or faulty. This notion of intuition sits nicely with how it is that mathematicians describe their experiences of learning mathematics. To the mathematician, the number 2 is very real, and very precise, or sharply delineated. Under normal circumstances, the mathematician will not mistake the number 9 for the number 2. Both are as real to the mathematician as tables and chairs.

Let us look at some arguments against the realist, in order to deepen our understanding. We shall turn to stronger arguments later, especially in Chapter 5. The problem with the realist account is that this notion of mathematical intuition is somewhat mysterious. For example, we might ask: what happens when someone has incorrect intuitions? Can we even make sense of the notion of "incorrect intuition"? The realist will say that, of course, we can have incorrect intuitions, whenever we make incorrect guesses about mathematics. We use reason to correct the mistakes. Can we have conflicting but legitimate intuitions? The realist will retort that while it might appear that there are disputes of this sort in mathematics, at the end of the day there is convergence on the views. This happens when we finally agree on a founding discipline. Convergence only happens after much argument, that is, after reason has had a chance to marshal our mathematical intuitions and perceptions. The process is piecemeal. For example, mathematicians all agree on the truths of the first four postulates of Euclidean geometry. This would be miraculous if we all had deeply felt divergent intuitions. This convergence confirms the hypothesis that the mathematicians are all perceiving the same things. These things are mathematical. The objects explain the convergence. Convergence is arrived at through reasoning. We have not yet worked out the absolute truth about Euclid's fifth postulate.

Unfortunately, the mystery of how it is that we come to know mathematical truths has not been solved. For, we might also ask where the intuition originates. Do we all have it from birth, or do some people have mathematical intuition, and some do not, a little like "extrasensory perception"? Certainly, this is what it feels like to a student who is struggling in a mathematics class. It seems as though some people just have a clear picture of what is going on, and that he just lacks this vision. The realist has an answer to this too: we all have mathematical intuition. It takes a good teacher, and one who manages the student's psychology, to draw out the student's intuition and perception. A good teacher can bring the student to understand, or perceive, the mathematical truths.

The realist has a strong position, since he can answer these difficult questions. So let us return to the notion of infinity, since it was already exercising the philosophers of ancient Greece. Plato himself has little to say about infinity, except in a spiritual sense. But we can imagine what Plato would say if he were interested in infinity, namely, that the realm of Forms contains an infinite set, say, the set of natural numbers. It might even contain several infinite sets: the set of natural numbers, the set of negative numbers, the set of fractions and so on. Cantor's "paradise" is related in this way to Plato's "heaven"! Cantor's paradise contains all sorts of infinite sets. Indeed, the modern mathematician who is comfortable with the notion of an infinite set tends to be a realist.

Twentieth-century realists in mathematics do say things about the infinite. What they say is that they accept the notion of the actual infinite. That is, there is some sense in which the truth or falsity of claims concerning infinite sets is independent of us, and in this sense objective. Moreover, there is an infinite number of infinite sets: all the ones in Cantor's paradise; sets of infinite ordinals and sets corresponding to the different cardinal measures of "size of set". Realists will usually fall short of insisting on the existence of a realm of objects that make the truths of mathematics true, despite the talk of paradise. On the other hand, the realist will be convinced that he discovers the truths of mathematics. "Discover" is, of course, used metaphorically, so does not need a realm of objects (as in Plato's heaven). We "discover" by means of our rational investigation.

"Discover" is to be understood as contrasted to "created". This can be a bit confusing in the literature, since realist mathematicians often use the word "create" or "construct" to refer to a procedure for deriving sets. When the realist creates a set, he "creates" in the sense of "following certain rules of the theory", where the emphasis is on "following". In contrast, the anti-realist (often called a "constructivist") thinks that we create mathematics, in the sense that the mathematics is "of our making", with the emphasis on "our". In set-theoretic realism, the axioms allow us to construct new sets from old ones in a very liberal way, compared to what the constructivists allow. The set-theoretic realist uses classical logic to construct new sets, and the reasoning used is of the form: *if this already exists, then this other thing must exist as well*. So, the notion of "construction" is that of set-theoretic construction, where the construction is simply allowed by the axioms of set theory. To make this clearer, it might be useful to look at the contrasting notion. The notion of construction, used by anti-realist philosophers of mathematics, is that of producing in a step-wise manner. We construct piece by piece. The anti-realist "creates" in the sense of following the rules in our minds. The anti-realist does not say that she "follows rules that we have *discovered*". This is because the rules are created by us. A consequence of this anti-realist thinking is that we are not allowed simply to gather things together in one fell swoop. In particular, the idea of just "taking the powerset" of a set is not allowed. For the constructivist this is a huge illegitimate leap, not a small allowed step. What the anti-realist, or constructivist, does allow is the construction of the powerset of a finite set, step by step. We have to be given a procedure for finding a first member of the powerset of a set, and then for finding a second member, and so on. We cannot just take the powerset, and move on from there, by making some further construction (in the realist sense) or comparing that powerset with another, or whatever we wish to do next. We shall discuss this in depth when we examine the anti-realist (constructivist) philosophies of mathematics.

For the realist, the repugnant part of anti-realism in mathematics is that if we simply go about creating mathematics, this makes mathematics sound like a fiction that we make up as we go along. This does not adequately account for the very stringent and rigorous practice of mathematics, or the strong conviction we have in mathematical truths. It dilutes any claims about discovering eternal truths.

Usually, the modern realist philosophy of mathematics takes set theory (instead of Plato's Euclidean geometry) as central to mathematics.[11] This is rather clever, because it allows the realist to be ontologically modest. Set theory only makes a very simple initial ontological claim: that, independent of us, there exists the empty set. Once we have this then we can develop, through the axioms of set theory, the whole set-theoretic hierarchy, which includes the finite sets and infinite sets. This is vast. The simplicity of the initial ontological commitment is what makes set theory "pure", very abstract and ontologically modest, at least *ab initio*.

Let us dwell a little on the axioms that perform the miracle of hoisting us from the modest ontological presupposition to a whole universe of sets arranged in a hierarchy. Set theory has only one pair of unexplained (called "primitive") notions. These are the notion of "a set" and that of "set membership". A set is a collection of objects. As soon as we think about sets, we learn quickly that there are laws governing them. These are the axioms of set theory. The axioms stipulate the existence of the empty set, and then tell us how to form a set from an already given set. In this sense we say that we construct the set-theoretic universe from the empty set, using the notion of membership of a set.

The notion of construction allowed by the realist is what gives set theory its great power to incorporate other parts of mathematics. In fact, most of mathematics can be faithfully redescribed by classical (realist) set theory. More precisely, we can translate other mathematical theories – such as group theory, analysis, calculus, arithmetic, geometry and so on – into the language of set theory. Arithmetic is another sub-part, and so on. In this sense, we might say that set theory is a "big" theory. Sometimes, we say that set theory is a founding discipline. So set-theoretic realism is also a reductionist account of mathematics, because of its ability to absorb other branches of mathematics. Most philosophies of mathematics are reductionist or foundationalist in some sense. The notable exception is structuralism, which we shall explore in Chapter 4.

We should pause here to say something about classical logic. Characteristics of classical logic (as opposed to non-classical logics) include:

(i) the law of bivalence;
(ii) the law of excluded middle;
(iii) the free use of *reductio ad absurdum* arguments; and
(iv) an ontologically significant reading of the existential quantifier.

We shall explain these briefly since some of them will come up later. Take this as a first exposure to the concepts. What we conclude about use of a classical logic is what is important here, not how we can spot one.

(i) The law of bivalence means that there are only two truth-values: true and false. In contrast, a trivalent logic will have three truth-values: true, false and undecided (or unknown). In a bivalent logic, every well-formed formula[12] will be either true or false. The truth or falsity might be independent of our ability to prove or know the truth or falsity of a claim.
(ii) The law of excluded middle is very similar to the law of bivalence. In classical logic the difference is quite subtle, for it is not a difference in what is being referred to, but in realm of application. The law of excluded middle says that given a well-formed formula, either it, or its negation, holds, and nothing else.

Strictly speaking, the law of excluded middle is syntactic, whereas the law of bivalence is semantic. That is, under the law of excluded middle, it is not possible for a well-formed formula and its negation to hold in the same theory. It is also not possible for a well-formed formula and its negation *not* to hold.

"Hold" means "is in the language of, and is consistent with, the theory", and not "can be proved". In a non-classical logic, either (i) or (ii) or both of these might be denied.

(iii) *Reductio ad absurdum* arguments are ones that start by denying what one wants to prove. We then prove a contradiction from this "denied" idea and more reasonable ideas in one's theory, showing that we were wrong in denying what we wanted to prove. We then appeal to the law of excluded middle and say that what we wanted to prove must be right after all. We saw an example of this structure of proof in Cantor's diagonal argument, which showed that the set of real numbers was strictly greater than the size $\aleph_0$.

(iv) The existential quantifier belongs to first-order logic and higher-order logics. It is symbolized by "$\exists$" and is read as "there exists" or "some". When a logic has an existential quantifier, it becomes much more powerful, pronouncing on what exists, and what does not exist (according to the theory). Some non-classical logics will read $\exists$ as "some" and attribute no ontological import to it; a separate statement is needed for existence.

Putting (iii) and (iv) together we have a type of proof that non-classical logicians object to: "purely existential proofs". These are proofs through a *reductio ad absurdum* argument that conclude with the existence of some object. If we think about it, what these proofs do is start by saying "suppose the object in question does not exist". We prove a contradiction from this, thus proving, classically, that the object must exist. For the realist classical logic is fine since we use the logic to reason about objects that are independent of us.

To summarize, the modern philosophers of mathematics who are closest to Platonism are the set-theoretic realists. A set-theoretic realist thinks of set theory as true and independent of us. Furthermore, it is the essence of mathematics because other parts of mathematics are reducible to it. The ontology of set theory is vast, in the sense that there are many sets. But it is pure, and perfectly abstract, in the sense of being constructed from the empty set. The rest of the hierarchy is constructed out of this. So it is almost as though the set-theoretic hierarchy is constructed *ex nihilo* (from nothing at all). That is the ontology of set-theoretic realism. What about the epistemology? Many set theorists describe their experience of "doing" set theory in terms of seeing or perceiving sets, or in terms of intuiting sets. The realist set theorists are entirely convinced of the veracity of their intuitions or their perceptions. They use classical logic to reason about these objects. One notable mathematician who discusses this in print is Gödel.

4. Kurt Gödel

Gödel (1906–78) is probably the most famous mathematician to express himself explicitly as a platonist. In a famous passage in his paper "What is Cantor's Continuum Problem?" he writes:

> But despite their remoteness from sense experience, we do have something *like a perception* also of the objects of set theory, as is seen from the fact that the axioms *force themselves upon us as being true*. I don't see any reason why we should have less confidence in this kind of *perception*, i.e., in *mathematical intuition*, than in sense perception, which induces us to build up physical theories and to expect that future sense perceptions will agree with them. (1983b: 483–4, emphasis added)

This quotation very much expresses the realist view. The epistemological account of mathematics is that we perceive axioms (of set theory) by means of a special sort of intuition. The ontology of mathematics is the set-theoretic hierarchy. The logic is classical logic: we construct (the hierarchy) in the sense of the logic allowing for the existence of new entities given the old ones.

The realist position is a little sneaky, or quite brilliant, depending on how we look at it. The realist relegates much of the philosophical work to the logic that is adopted to underpin the mathematics. This is clever because logic is thought of as entirely primitive, universal and unified, so we are reluctant to argue against it.[13] The anti-realist shows great courage by arguing that adherence to classical logic will get us into trouble if we are not careful. For example, Bertrand Russell (1872–1970), who made one of his most important contributions to the philosophy of mathematics thirty years before Gödel's most important contribution, reacted to the set-theoretic paradoxes by warning against the use of "impredicative definitions" in mathematics.[14] Impredicative definitions are ones that refer to themselves (and they might lead to paradox). An example is that "a great general" is defined as a military person of the rank of "general" and who is great. The definition is circular. If we do not understand the terms "general" and "great" in advance, then the definition is not helpful. In mathematics, this sort of definition is doubly unhelpful because we are often breaking new ground, so we do not really know in advance where we are going. Russell concluded that our definitions have to be precise, and "predicative". Definitions have to define a term using wholly different objects or concepts; or so Russell thought.

Gödel argued against Russell's counsel. He argues that if we are realists, then there is nothing fundamentally wrong with impredicative definitions. They simply partly tell us about something. But, that something is already known to us through our intuition. "If, however, it is a question of objects that *exist independently of our constructions*, then there is nothing in the least absurd about *the existence of totalities containing members*, which can be described (i.e., uniquely characterised) only by reference to this totality" (Gödel 1983a: 456, emphasis added).

There is nothing wrong with using classical logic and impredicative definitions when the sets exist independently of us. Let us explore this further. In the quotation, Gödel is reacting to Russell's warnings against the use of impredicative definitions. In doing so, Gödel reveals an important presupposition that is built into the logic used in set theory. The logic being used is classical logic. In classical logic, definitions are thought of as revealing our attempts to refer to objects (which exist independently of us and our attempts to refer to them). A definition is good or bad according to how well it refers. Partial, or imprecise, reference is not so good.

In contrast, in intuitionist or constructivist logics, how we formulate a definition or how we construct an object by means of a definition is extremely important. This is because the objects are constructed and not independent of us. If we are not careful in our constructions, then we risk getting into trouble, in the form of generating an inconsistent theory. If we find an inconsistency, then the whole system falls. Paradoxes indicate inconsistency. Sometimes we can explain them away, and show that they do not really show an inconsistency. But we have to do quite a lot of work to do this, and the results are not always satisfactory. So, we have to get it right in the first place because we are determining an object; creating it, in a sense. If our definitions do not uniquely characterize an object, we are not entitled to discuss the object.

To summarize, Gödel's remark says that since the objects of mathematics, such as sets of numbers, exist independently of us, it does not matter how we refer to them, provided our language is sufficiently clear to avoid generating outright contradictions. Definitions are supplemented by our intuitions; they hone our intuitions. Gödel's diagnosis, when we do end up with a contradiction, is that we have used poor language or are confused in some way. In other words, *our* name for the sets or objects is what is muddled, not the objects. The objects are well organized, quite independently of us.

In contrast, if one is not a realist or a platonist, then impredicative definitions are a problem. This is because the objects studied by mathematicians are not independent of the names we give them. Gödel's more philosophical material has, for the most part, only recently been made readily available. In the texts that were available for a long time, he did not say much at all about mathematical intuition, and how that is supposed to reveal mathematical truths to us. One philosopher who made a valiant attempt to explain the mechanism of mathematical intuition is Maddy. We shall now turn to her account, as a possible philosophical supplement to Gödel's platonism. It will take some time for philosophers to digest Gödel's complete works; we can anticipate more insights into Gödel's thinking surfacing in the philosophy of mathematics.

5. Penelope Maddy

Maddy picks up the thread where Gödel left off. In "Perception and Mathematical Intuition" (1996) she gives philosophical arguments and philosophical support for Gödel's platonism: "Taking Gödel as a starting point, I will try to sketch an account of perception and intuition which will ... provide an account of set theoretic realism" (*ibid.*: 116).[15] The main philosophical problem with the position of platonism or realism is the epistemic problem: that of explaining what perception or intuition consists in; how it is possible that we should accurately detect whatever it is we are realists about. Or, if we are platonists (or Platonists), how it is that we should have insight into Plato's heaven, and how we can feel at all confident that we have accurate, veridical insight/perception/intuition of this reality/heaven.

Maddy is interested in interpreting what it is that Gödel means when he says that mathematical intuition is something "like sense perception". To give scientific and philosophical credence to the view that mathematical intuition is a type of sense perception, there has to be a physical mechanism in the human body to account for the perceiving. This is what it takes to give the right sort of causal story to explain mathematical intuition.

What does "right sort of causal story" mean? At the time of writing the article, and among her philosophical colleagues, two points were taken for granted: that perception has to be explained causally, and therefore physically;[16] and that what the mathematicians say about their experience of mathematics is what the philosophers should work to support. Stewart Shapiro calls this position "mathematics first, philosophy second" (2000: 7–20). The idea is that if philosophers want to know what mathematics is, they had better consult with the mathematicians, since they are the most qualified to tell the philosophers. Similarly, if mathematicians' experience of mathematics is best described by one philosophical position rather than another, then the philosopher had better work on that philosophical position and not on another position. The philosopher plays second fiddle to the mathematician. Maddy lists this as an advantage of her position. Maddy's set-theoretic realism "squares with the pre-philosophical views of most working mathematicians" (1996: 115). This orientation is not universally favoured by philosophers; it is favoured by Maddy, Shapiro, W. V. Quine and possibly Wittgenstein.

The contrasting view would be one of philosophy first, where the philosopher works out what is the most defensible position in the philosophy of mathematics, and then tells the mathematician what mathematics is. Note that this orientation of "philosophy first, mathematics second" is potentially revisionary of mathematics. That is, the philosopher with this orientation will have no qualms telling the mathematician that certain moves in proofs should no longer be allowed, or that certain branches of mathematics are pure fantasy, bordering on the irrational. We shall see some of this in Chapter 5 when we look at constructivism.

Returning to Maddy, she is of the "mathematics first, philosophy second" persuasion. She also thinks that intuitive perception has to be given a causal, and therefore physical, account.

> Set theoretic realism is a view whose main tenets are that sets exist independently of human thought, and that set theory is the science of these entities. (1996: 114)

> On the realistic assumption that they [sets] do exist, I will try to show how we can refer to and know about them. Since accusations to the contrary are most often based on *causal theories of reference and knowledge, I will use these theories as starting points.* (*Ibid.*: 116, original emphasis)

Maddy describes some of the clearest and most damning current objections to set-theoretic realism, which states that the intuitive faculty cannot be explained in causal physical terms and therefore there cannot be such a faculty.

Maddy then goes on to describe these causal theories of perception and adapts them to set theory, arguing that it is no more mysterious that we should perceive that there is a set of a dozen eggs in front of us, than it is to perceive that there is a chair in front of us. In fact, the mathematician is someone who will perceive that there is the singleton set of a chair in front of us.

It sounds extravagant to say that we perceive sets, or any other mathematical objects for that matter, however, we should be aware that Maddy draws a careful and subtle distinction between "physical seeing" and "perception". The distinction does much of the work in her theory. "Seeing" is a purely mechanical process entirely explainable in terms of the function of the eye, light, and the wavelengths of colours around us. In contrast, perceiving is interpretative. That is, when we perceive, we see but also look for some features and ignore others. When we perceive, we see selectively. Selection is a mental process. For example, we individuate objects from the mess of colour splotches. The individuation (telling which object is separate from which other object) is an act of perceiving. More importantly, perceiving is partly a linguistic activity. We look for things we have words for. We perceive a chair, but we see a brown splotch with some straight edges. Similarly, we perceive sets; we do not see them. They form part of our interpretation of the world. We learn to perceive and in learning to perceive we first have to have a concept of the object we are perceiving. When a child is having objects pointed out, she is learning to pick out instances of the concept named by the noun being used to name the type of object. To learn to perceive sets, we have to learn some set theory. This is how Maddy interprets Gödel. When Gödel says that we "perceive the objects of set theory" (1983b: 483), what he means is that we perceive in the sense of "selectively see". The selection is informed by our learning of set theory, which apprises us of the concept of set. We are then sensitive to the notion, and can then perceive sets.

But then we have a problem. The seeing part is entirely mechanical and causal. This can be given the scientifically reputable explanation the critics of mathematical platonism seem to require. However, the work of mathematical intuition is not being done purely by the mechanism of seeing; rather, it is being done in the mind, allowing us to pick out certain "set" features. The picking out requires learning concepts and interpreting the splotches we see. Now, on the one hand we have an explanation as to why it is that people without mathematical training fail to perceive objects organized in sets; on the other hand the reputable causal part of the explanation of perception is doing no work. The justification for set-theoretic realism is being given by the mathematician's training in set theory. This is what gives her the ability to interpret what she perceives. The mathematician interprets what she perceives in terms of set theory. This is circular. There is no guarantee that sets lie in the physical world independent of us. On the contrary, sets of objects seem to be highly dependent on our training in set theory. This would suggest that sets are created by us, and therefore depend on us, and this is at odds with realist thinking.

Moreover, there is no way that Maddy can account for the existence of large cardinal sets, since we neither see nor perceive these. Yet set theory is committed to the existence of sets of different infinite cardinality. The problem that Maddy's account encounters with this argument is to make perception supervene on seeing.[17] We cannot perceive without there being some physical seeing. Since we never see an infinite set (since there are only a finite number of physical objects in front of us at any one time), we also cannot perceive an infinite set.

We could counter, on Maddy's behalf, that there is some sense in which we see an infinite set: when we think of anything continuous, for example, we might stand on a spot, and perceive that there are an infinite number of directions one could (in principle) face from that spot. The problem with this move is that it is not clear that we can perceive anything of the sort. Again, the interesting aspect of perceiving will depend on our education in set theory; we cannot then appeal to perceptual intuition to explain our development of set theory, for this would be a circular explanation. Furthermore, even if we can make sense of this story of perceiving an infinite set, this is not enough to explain how we might differentiate between a set of size of $\aleph_1$ and one of size $\aleph_0$, let alone any set of greater cardinality, and yet such sets are part of the classical mathematician's training, which should allow her to perceive sets of these sizes.

Maddy's valiant attempt to defend set-theoretic realism along the lines of a physical and causal explanation of our intuiting sets fails. Realism seems plausible, and echoes the experience mathematicians have of mathematical objects; however, it is not a watertight position.

Before losing all hope of defending realism we should add a few notes. One is that Maddy has gone on, since the quoted article, to modify her position. This shows that her views are developing. Maddy now defends a naturalist position in the philosophy of mathematics. So it is unfair to dismiss Maddy on the grounds of the quoted article. The article was mentioned here as a brave attempt to defend a very plausible position, and to show how difficult it is to do so. Once we have realized this, a change of tack is not absurd. If we look back, it is only when we insist that the only legitimate justification for mathematical intuition is an explanation along physical and causal lines that we end up in this mess. It is only if we think that perception supervenes on a physical account of seeing that we feel the pressure to ensure that sets get their pedigree from physical causal accounts of seeing in just that same way as the objects studied by the physicists do. As philosophers of mathematics, we can resist these demands made on which types of explanation are good and which are not.

A last note is that it is quite true that many mathematicians are attracted to realism in mathematics. Again, if one is persuaded by the "mathematics first, philosophy second" orientation, then one will want to take heed of what the mathematicians tell us. However, it should be noted that this is partly a geographical and historical prejudice. Throughout the twentieth century many Russian, Central European and Eastern European mathematicians have been very aware of, and sensitive towards, constructivist views. They show a clear preference for constructive proofs, and many are steeped in an anti-realist orientation towards mathematics.[18]

To set the stage for the remaining chapters, let us examine closely the general problems with set-theoretic realism, since this philosophical position is central to the philosophy of mathematics. Along with set-theoretic realism, we have a strong endorsement of the notion of actual infinity, together with the developed set-theoretic notions of ordinal and cardinal infinite numbers. All sorts of infinite sets exist for the realist, and this is very attractive to many mathematicians.

6. General problems with set-theoretic realism

Present-day realists tend to be set-theoretic realists, that is, they tend to take set theory as the founding discipline of mathematics. There are two easy problems with set-theoretic realism, and two hard problems, and the problems can be revamped to apply to other sorts of realism. The problems come in two areas: ontology and epistemology. The problems concerning ontology have to do with (i) the number of objects that exist according to the theory and (ii) which objects there are. The epistemological problems concern (a) how we justify that said objects exist and (b) how we apprehend them. There are quite good answers for (i) and (a); (ii) and (b) are more difficult to answer.

Let us deal with the easy questions. The ontological question about the number of mathematical objects (i), generates a traditional worry in philosophy, often referred to as "Ockham's razor". William of Ockham (1285–1347) complained about certain scholastic views about angels that concluded that since angels are ethereal they do not take up any space. It followed that an infinite number of them could dance on the head of a pin, and so there was nothing (spatio-physical) preventing the existence of an infinite number of angels "located" in a very small space. Ockham thought that this was too much: there was no reason to think that there was an infinite number of angels. Ockham issued a wise principle to guide our metaphysical thinking: a theory should not postulate more objects than are necessary for the theory.

In philosophical circles, the Ockham's razor principle is part of the antidote to philosophical queasiness that is felt when confronted with a large number of abstract objects. When evoking Ockham's razor, the suspicion is that we can make do with fewer abstract objects, or none at all. In other words, it is considered to be a fault of a theory to postulate a large number of abstract objects. When we heed the principle, we cast around for a theory that has fewer abstract objects, but that explains the same phenomena equally well.

The "how many" problem comes from a sense of discomfort. The whole set-theoretic universe includes infinite numbers of infinite sets of abstract objects. Do we need them all? The worry can be answered by drawing attention to some considerations we made earlier. The set-theoretic hierarchy is constructed out of the empty set. The empty set is not vast; it will neither take up a lot of room, nor are there many of them from which we are constructing the set-theoretic universe. After all, we use the direct article when discussing "the" empty set. If the empty set is any number of things, it is either zero things or one thing. So Ockham's razor cannot do much cutting here.

One might object by drawing attention to the axiom of infinity, which stipulates that there is an infinite set. In fact, there are a number of these axioms, one for each size of infinity we wish to include in our theory, which we cannot construct using the powerset operation (provided the new axiom is consistent with the other axioms). In principle, we could have an infinite number of these, since there is no reason to think that there is any end to adding axioms of infinity. The retort made by the set-theoretic realist is to point out that even these infinite sets are made up of the empty set, so they only give the impression of committing us to the existence of an infinite number of new objects.

The first epistemological question (a) also admits of a reasonable answer. The question was: how do we know that the ontology of set theory exists? This boils down to the question: how do we know that the empty set exists? We certainly talk about there not being anything. For example, one might say that there are no unicorns or, more mundanely, there are no eggs in the refrigerator. It is easy now to get into a muddle, because the grammar treats "nothing" as an object. There is nothing in the bucket. However, "nothing" is not a physical object, despite the fact that we sometimes attribute location to "it": there is nothing *in* the bucket. Our concepts and our grammar allow such locutions. Our concepts and grammar do lead us into a muddle if we investigate our use of the concept of "nothing".[19] The set theorist cleverly points out that, in set theory, we do not have an analysis of the empty set, or of "nothing". Instead, we introduce it as a primitive. We know about it mathematically through the axioms of set theory. Further analysis of the notion of the empty set belongs to metaphysics. In set theory we do not get muddled. Furthermore, we are quite certain that set theory is consistent.[20] Proofs are epistemologically very secure. How do I know some truth about set theory? Because I have a proof. Proof is evidence by reasoning, and it is a sort of ultimate evidence. That is what we mean when we say that there is a proof that something exists. It is no longer up for question. There can be no counter-evidence if we have a proof. This is about all that can be said to answer epistemological question (a).

What of ontological question (ii) (which objects there are) and epistemological question (b) (how we apprehend the objects)? Both problems are exacerbated by the presence of competing set theories, which disagree over the truth or falsity of certain axioms. There are two sorts of situation where we have rival theories. One is where we have slightly different basic axioms. This makes for two rival set theories that will disagree on certain issues. Examples are Zermelo–Fraenkel set theory and Gödel–Bernays set theory. The two theories disagree over the notions of ordinal construction and over the notion of class, among other things.

The other way in which we can have two opposing set theories concerns what are called "independent" axioms. That is, we keep the basic axioms fixed, but then there is a question left pending about further axioms. There are some, non-basic, axioms that can be attached to the set theory or not. This is optional. An example is the axiom of choice. There are different versions of this, but basically it says that there is a way of picking out one member of every set that is a member of some larger set. For example, consider a set whose members are all pairs of things: pairs of socks, pairs of earrings, pairs of gloves, pairs of railway tracks and so on. The axiom of choice tells us that there is a mathematical way, guaranteed by the axiom, to pick out one of each. The way has to be mathematical; it cannot just be something like "pick out the first you come to" since there are often an infinite number, and the objects are abstract, so there is no "first you come to". Take a more abstract example. Consider the set of arbitrary pairs of numbers (where each member of the pair is different from the other). Each member of the set is a pair. The axiom of choice says that there exists a mathematical way of selecting one of each of the pairs. In this abstract example we can come up with a choice function: we could say, for example, take the lesser number of each pair. This is a "nice" example of the axiom of choice. The axiom guaranteed that there was a choice function. We were then able to specify such a choice function. However, in mathematics we cannot always specify a choice function. The axiom simply tells us that a choice function exists. It does not tell us how to specify the function. It is these sorts of cases that lead some mathematicians to object to the axiom of choice, saying it is simply not true, unless one gives a method for picking out one object from each set. Other mathematicians endorse the axiom. Neither Zermelo–Fraenkel set theory nor Gödel–Bernays set theory tell us one way or the other whether the axiom is true. The attaching of the axiom of choice to either theory is consistent with the rest of the theory. The denial of the axiom is consistent with the rest of the theory. When we have this situation, we say that the axiom of choice is "independent" of the set theory. We have two rival theories: Zermelo–Fraenkel set theory with the axioms of choice attached, and Zermelo–Fraenkel set theory without the axiom of choice attached. Axioms of infinity are independent of set theories too. The outcome of all this discussion is that there are rival set theories, and it is not at all clear which is the "true" one.

This interferes with how we are to know about the basic components of set theory. For example, we are told that we are to take the notion of set as primitive, or undefined. The problem then is that the only purchase we can get on the notion of set is through the axioms. We say that the axioms give an "implicit definition" of the notion of set. Since competing set theories have different sets of axioms, there are competing notions of set. The problem with this is that there does not seem to be a good reason to choose one implicit conception over the other, so we still do not really know what the truths of mathematics are in the final analysis.

Rescinding the "primitive notion of set", we might say that our purchase on the concept of set does not come from the axioms of set theory, but rather from intuition, or from some sort of perception. We saw that Maddy was the person to most fully develop this view. She does so with mathematicians such as Gödel very much in mind, and in this sense, represents his views to the philosophical community. Since he is one of the most respected mathematicians of the twentieth century, his philosophical views concerning mathematics, and any position developing his views, should be carefully scrutinized. Unfortunately, as we saw, either Maddy is not a very good representative of the modern position of mathematical realism, or it is simply not a philosophical position that bears much scrutiny. At the end of the analysis of Maddy's development of Gödel's position, we do not have a good account of either which objects there are (how to best characterize sets) or how it is that we apprehend these. We do not apprehend these through perception, and we cannot give philosophically sound reasons to choose to study the notion of set through one set theory as opposed to a rival theory. That is, our intuitions do not guide us all to the same choice of founding discipline in mathematics, and therefore intuition is not a reliable guide to determining which mathematical objects there are.

To be very precise about where we are in assessing the strength of the realist position, our consideration of competing set theories is not enough to disprove realism. For, a realist could simply insist that his choice of founding discipline is correct. If that choice rests on a gut feeling, or intuition, then we have a philosophical problem. This is because realism about mathematics, sets or anything else insists on the mind-independence of mathematics, sets or whatever. This comes at a price. The price, which is a brute fact that follows from mind-independence, is that it is possible that we have hitherto failed to detect or latch on to these entities. We might think that this is highly unlikely or, at least, that it would be very unfortunate if it were the case. Nevertheless, it is perfectly consistent with set-theoretic realism that there should be competing notions of set. Moreover, they might all be wrong.

While consideration of rival set theories does not disprove set-theoretic realism, it certainly discredits it. One has to ask what the point is of even saying that there must be sets, if we have no solid way of knowing that we are dealing with the truth of the matter. Our feelings or intuitions are not a good guide because they lead different mathematicians to mutually incompatible conclusions and notions of set.

An intelligent and plausible answer to epistemological question (b), and thereby ontological question (ii), is advanced by Eckehart Köhler (2000). He argues that there is an intuitive faculty, which he calls "rational intuition", or "reasoned intuition", which is what mathematicians use to perceive the set-theoretic hierarchy, and do their mathematics. This is not a faculty of perception in Maddy's sense.[21] It is not causal, and does not supervene on seeing. Instead, it is based simply on our capacity to reason. We are all able to reason. Thus, we are all capable of abstract thinking. It is our reasoning intuition that guarantees that we are able to do this at all.[22] Rational intuition is what allows us to abstract from what we are thinking about. Moreover, our rational intuition allows us to reason about abstract objects such as mathematical entities. Köhler's thesis is closely supported by passages in Gödel, where he mentions specifically a notion of rational intuition, which is different from sense perception. Köhler is a better representative of Gödel's realism than Maddy.

There is a difficulty. People reason in different ways, so there is not one thing that is our reasoning intuition. However, following Gödel and Köhler, we can restate the position in a more sophisticated way. Different people have reasoning intuition developed to different degrees. This is why we say that one person has greater reasoning powers than another, for example. This is also why, when we point out a fault in reasoning to someone, she is able to recognize the fault. Interestingly, when confronted by someone who does not recognize, or refuses to concede that he recognizes a fault in reasoning, we tend to dismiss the person as irrational. Some people have a better developed sense of reasoning than other people, just as some people have a better sense of geometry or arithmetic than other people.[23] In fact, we can be trained to reason better. This is why we teach classes in basic logic and critical thinking. These classes develop our reasoning intuition. So far, so good, but what about rival set theories? It is perfectly consistent with Köhler's view that we have not yet worked out which is the true set theory. Our collective reasoning powers are not sufficiently developed to have settled on one set theory. Set theory is rather young after all. However, it is also consistent with Köhler's view that we shall never settle the matter. That is, our intuitions might not be strong enough, or maybe never will be strong enough, to be able to tell which is the real set theory. We can be optimists or pessimists.[24] Gödel was an optimist; Köhler is a Gödelian optimist.

We should now step back to assess the view. The attraction of Köhler's view consists in its responding well to the phenomenology of mathematics, that is, to how it is that mathematicians describe their experience of mathematics. Köhler's development allows for reasoning intuition leading to disagreements among mathematicians. So the view offers a good diagnosis of, and explanation for, mathematical practice. The problem is that it does not get us closer to what we want. We cannot explain the mechanisms of this intuition. It would be circular to appeal to formal logic or set theory. It would be hopeless to appeal to a physical mechanism, as Maddy does. We shall leave it up to the reader to decide whether such a "mechanistic" explanation of intuition is needed for a realist theory. Insisting on a mechanistic explanation, grounded in the physical, leads to empiricism or naturalism. If we do not insist on a mechanistic explanation of reasoned intuition, then we are left with an attitude of optimism or pessimism concerning our reasoning powers.

So, while our intuitions might be all we have to go on, and while the notion of intuition might well describe and capture the experience or phenomenon of practising mathematics, it is not an entirely satisfactory theory from the philosophical point of view.

7. Conclusion

As we have seen, the philosophical position called mathematical Platonism is very ancient. Arguably, it is the oldest position in the philosophy of mathematics. More precisely, Plato was the first[25] to write seriously about the nature of mathematical knowledge, as opposed to other sorts of knowledge. Our knowledge of mathematics seems to be entirely solid, in the sense of indubitable. We acquire mathematical knowledge through proofs and demonstrations that have a strong force to them. Proofs are rationally compelling. Interestingly, the proofs of geometry, for example, do not rely on the accuracy of the diagram we use, or particular examples. We seem, for Plato, to be investigating perfect and general objects when we make proofs based on the axioms of Euclidean geometry. For Plato, these perfect objects have to exist independently of us, and since they have to exist, then they have to exist "somewhere". They exist in Plato's heaven.

The realist in mathematics falls short of this strong ontological commitment to a Platonic heaven, by not postulating a heaven of perfect objects, but nonetheless claiming that mathematical truth is independent of the human mind. That is, the truths of mathematics are timeless. This can be understood in the sense that if all rational beings, that is, beings practising mathematics, were to die, the mathematical truths would still be there, as they were before we did any mathematics. Human beings are fortunate enough to *discover* the truths of mathematics. Furthermore, we think ourselves lucky because we believe that we have a fairly accurate picture of what the mathematical truths are. That is, our mathematical intuition and our way of confirming this through proof are quite good at getting it right.

Moreover, this is confirmed in the descriptions many mathematicians give of their experience of doing mathematics. For many of them, mathematics is a process of discovery of quite independent truths. Few mathematicians use the language of creating mathematical entities, although they will talk of constructions, so there is some ambiguity even at the level of candid language.

The philosophers who give the most depth to the notion of intuition are Maddy and Köhler. Maddy gives a physical and causal theory of set-theoretic realism. In providing a philosophically worked-out account of mathematical intuition, she confronts the greatest problem facing the mathematical realist, namely, why we are so certain about mathematical truths, and why it is that we think that they are independent of us, and timeless.

Köhler's development of the notion of mathematical intuition is much more subtle. He postulates a rational intuition, to be understood as a reasoning faculty which we, as human beings, have. Köhler's rational intuition is the faculty by which we apprehend mathematical truths. It is not perceptual, as it is with Maddy. Rational intuition does not fit well with a rigid causal theory of knowledge.[26] This is a distinct advantage of Köhler's realism over Maddy's.

There is no denying that we feel certain when we give a proof in mathematics. We do seem to be making discoveries when we learn mathematics. We usually do not feel that we are creating new truths that did not previously exist and that are not responsible to some independent reality. This is because there are constraints in developing mathematical theories. It is not the case that "anything goes" in mathematics. The rigour of mathematics fits well with the notion of discovery. If we were simply creating mathematics, then it would not be clear why it is that we have to be so rigorous about it.

The problem is that while many mathematicians have realist inclinations, it is not obvious how to defend mathematical realism philosophically. It is also clear that there are conflicting purported truths in mathematics in the sense that some sentences are true according to one theory and false according to another. For example, we might choose a sentence proclaiming the existence of infinite sets. These exist in some theories and not in others. So the sentence asserting the existence of an infinite set is true in one theory and false in another. How do we choose between competing theories? Take a poll? Trust the feelings of the greatest mathematicians? Assume that the "biggest" theory must be the true one? Take a theory that is common to all mathematical theories? None of these is an obvious and compelling choice over others. More importantly, none are philosophically persuasive.

8. Summary

We now have a default philosophy of mathematics. We also know that it engenders philosophical criticism. The important points to retain from this chapter are:

- the distinction between realism in truth-value and realism in ontology;
- realism about a subject, x, consists in thinking that the ontology and/or truth-values of the theory are mind-independent;
- the main problem with realism concerns the epistemology of the theory;
- the appeal of realism is that it accords with reports of mathematicians on how they view mathematics.

Chapter 3
Logicism

1. Introduction

Logicism was advocated by Richard Dedekind, developed by Gottlob Frege and extended by Bertrand Russell (1872–1970) together with Alfred North Whitehead (1861–1947). We shall not say anything about Dedekind here, but take up the philosophical position starting with Frege. Frege developed logicism through three works. The first, the *Begriffsschrift* (*Concept Script*) (1976),[1] first published in 1879, is a technical work, introducing the reader to a formal logical system. The second work, the *Grundlagen* (*Foundations of Arithmetic*) (1980a),[2] first published in 1884, is philosophical. The third, the *Grundgesetze der Arithmetik* (1980b),[3] was originally published in two volumes, in 1893 and 1903. It would also be translated as *Foundations of Arithmetic*, but these are formal foundations, not philosophical ones. Whitehead and Russell continued the logicist project, and published *Principia Mathematica* (1910–13) in three volumes. This is a technical work developing a formal theory of types, which, they argue, is pure logic. Russell also published more philosophical works: *The Principles of Mathematics* (1903) and *Introduction to Mathematical Philosophy* (1919).

The major philosophical question the logicist tries to answer is: what is the essence of mathematics? As the word "logicist" suggests, the answer is that mathematics, or part of it, is essentially logic. Logicism can be either a realist philosophy of mathematics or an anti-realist philosophy of mathematics. Frege was a realist. Frege's logicist believes that mathematical truths are independent of human beings. Frege's version of logicism is epistemologically realist. That is, human beings discover, or fail to discover, mathematical truths. Frege's logicism is realist in truth-value. Logical truth is independent of human beings. Frege's realism in truth-value is underpinned by his realism in ontology. That is, what accounts for the independence of the logical truths is the existence of logical objects.

The Fregean logicist differs from the Platonist, platonist or realist in how he attempts to vindicate his realist stance. The logicist seeks to support his realism by first showing that all, or part, of mathematics is really logic, and secondly supporting the claim that logic is objective. To show this, the logicist has to reduce all, or part, of mathematics to logic. The reason the locution "all or part" is being included above is that Frege believed that arithmetic is reducible to logic, but not necessarily other parts of mathematics. In particular, according to Frege, geometry is not reducible to logic.[4] Whitehead and Russell are more ambitious. They think that all of mathematics is reducible to logic. What is common to the two views is not only the reduction to logic but also the conviction that there is something deeply fundamental about logic; that logic occupies a privileged place, not only in mathematics, but in all our knowledge. More dramatically, logic is foundational to mathematics. Logic is more basic and universal than the rest of mathematics.

This is at odds with much current thinking about logic, where logic is thought of as one branch of mathematics among others. Logicians in mathematics departments think of themselves as studying a branch of mathematics, on a par with topologists, geometers, set theorists and so on. The thought that logic is more fundamental than the rest of mathematics is a philosophical position about epistemology. We shall discuss this shortly, and again in more depth in §3.

Whitehead and Russell's logicism differs from realism in that they do not believe that mathematical truths are independent of us. In some sense, we have to construct them and, moreover, we have to be very careful about how we construct them. Only some techniques are permissible, and these are logical techniques. The sense in which Whitehead and Russell are logicists is that they believe that mathematics is essentially logic, and they develop their formal theory of types in order to prove this, by showing that they can reproduce all of mathematics using their type theory.

Whitehead and Russell's logicism differs from Frege's in two respects. One is that Frege is a realist and Whitehead and Russell are not. The other respect in which the logicist positions differ is in their scope. Frege thought that only arithmetic and analysis are branches of logic. Whitehead and Russell think that the vast majority of mathematics is essentially logic. The realist/anti-realist divide is interesting in the case of logicism, for it does not concern infinity in the way discussed in Chapter 1. That is, we do not have a notion of actual infinity in the case of Frege, and a notion of potential infinity in the case of Whitehead and Russell. Instead, Frege, Whitehead and Russell all embrace the notion of actual infinity, and are at pains to ensure that they can capture infinite sets within their respective formal theories. This attitude towards infinity is something they share. Nevertheless, it is not really an important way of characterizing the logicist position. In order to be considered to be a logicist, it is not necessary that one endorse an actual notion of infinity. However, whether one does will have an effect on the scope of logic: on which parts of mathematics turn out to be essentially logic. The realism/anti-realism divide concerns the independence of mathematical truths from our abilities to conceive them. Simply put, for Frege we discover mathematical truths and for Whitehead and Russell we create them. Whitehead and Russell are particularly careful to avoid paradox, and consider the paradoxes to indicate that we create mathematical reality, since we have made the mistakes that led to the paradoxes. For this reason we have to be very careful in how it is that we justify the "truths". They should be justified by a completely reliable system, which will not lead to paradox, and this is their logical type theory.

If well demonstrated, logicism is a very appealing and strong position because it makes explicit use of a firm conviction that underpins much of our thinking. We think of disciplines of research and enquiry as arranged in a hierarchy. The most general of these is logic. Next down we have (the rest of) mathematics. Below this we have physics, then chemistry, then biology, then the social sciences and then it becomes quite sprawled and messy (Fig. 13). The truths of level 6 depend on those of level 5; the truths of level 5 depend on those of level 4; the truths of level 3 depend on those of level 2; and so on.

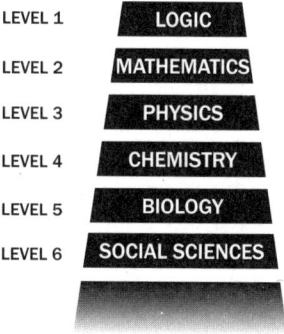

Figure 13

The thinking behind the level ordering in the hierarchy is that when we investigate questions in biology, for example, we have to assume the laws of chemistry but not those of economics. Another way of expressing this is that we could never find a principle of biology that contradicted something in physics or in chemistry. Since logic occupies the top of the hierarchy, logic is the most fundamental discipline. Moreover, the logicists believe that all, or part, of mathematics belongs to the most fundamental discipline: logic. This further move of reducing all, or part, of mathematics to logic accompanies the very strong conviction that the truths of mathematics, or arithmetic, are undeniable; 2 + 2 = 4 seems to be an absolute truth, not up for debate. In other words, the reason mathematical, or arithmetical, truths are not up for debate is that they are really logical truths, and there is no position from which we may question these. If we can show that mathematics, or arithmetic, is really logic, then we show how solid and unquestionable mathematical, or arithmetical, truths are. A contrary, and also very strong, conviction underpins the empiricist, fictionalist and naturalist positions. We shall see these in due course.

Frege did not think that all mathematical truths are essentially truths of logic. Rather, he thought that arithmetical truths are truths of logic, but that geometrical truths are not. This is because we cannot deny the truths of logic or arithmetic. In contrast, we can imagine denying the "truths" of Euclidean geometry. Frege writes: "For the purposes of conceptual thought, we can always assume the contrary of some or other of the geometrical axioms ..." (1980a: §14). Compare this to his thoughts about arithmetic: "[if we] try denying any one of them [the fundamental laws of logic or of number], [then] complete confusion ensues. Even to think at all seems no longer possible" (*ibid.*). It is quite a nice experiment to try to think of ways of denying the arithmetic fact that 8 + 765 = 773. An easy way is to just reinterpret what the symbols for the numbers stand for. For example, reinterpret 8 to mean 20, then "8 + 765" will equal 785. But this is cheating; it is changing our language, not the underlying truth. For the proper experiments we should keep the language fixed because we are interested in the underlying truth, not in the names for the numbers. If we keep the language fixed, then it is hard (read impossible) to deny that 8 + 765 = 773. This is because it is impossible to propose a viable alternative to arithmetic, where by "viable alternative" we mean a theory that is just as convincing and respectable as arithmetic.[5] The thought experiment reveals the deep convictions that motivate logicism, and so serve as a backdrop to the rest of the chapter.

In this chapter, we shall take a fairly in-depth look at Frege's development of logicism. We shall look at it from the technical point of view, and from the philosophical point of view. We shall then see what is wrong with the position, and then how we might react to the criticisms. We then look at Whitehead and Russell's logicism, and problems with this. Lastly we look at some modern developments in logicism.

2. Frege's logicism: technical accomplishments

In order to show that arithmetic is really logic, Frege developed a sophisticated system of logic. His *Grundgesetze* contains long proofs, in Frege's formal system, that the Peano/Dedekind axioms of arithmetic are really just theorems of logic; thus reducing arithmetic to logic. Russell then showed that Frege's system of logic suffers from a fatal flaw: it is inconsistent. Nevertheless, Frege's attempt at reducing arithmetic to logic should not be overlooked. We learn rich lessons in studying Frege's logicism.

There had been developments in logic between Aristotle and Frege. In ancient Greece alternatives to Aristotle's syllogistic reasoning were developed, but they did not survive into the Middle Ages. The medieval philosophers developed the syllogistic reasoning beyond Aristotle, and also experimented with a few other systems, but they were not well received by the universities, and were dropped. Logic and mathematics were perceived as being very different disciplines. Logic was a branch of philosophy, not of mathematics. This remained the norm until George Boole (1815–1864), and later Frege, decided to bring their mathematical understanding to bear in the realm of logic. Boole made some major advances in logic in his *Laws of Thought* (1854), in which he brings an algebraic approach to propositions and introduces the notion of a quantifier and a type of probabilistic reasoning. Boole's formal system was not taken up by philosophers, despite the fact that he makes a considerable effort to show the direct application to philosophical reasoning by translating arguments from Spinoza and Samuel Clarke into his formal notation, and exposing the formal reasoning.[6] Despite Boole's achievements, the system he advances is not sophisticated enough for Frege's purposes. Nor is the syllogistic reasoning. In particular, before 1879, logic was not powerful enough to derive the axioms and theorems of arithmetic. Thus, Frege was compelled to develop a new system.

Frege's formal system was a revolutionary advance in logic. He brought concepts from mathematics to bear in the analysis of sentences, and his judicious choice of notation and concepts borrowed from mathematics and linguistics made for a very powerful system. The advantages of a powerful system over a weak system are that in a powerful system we can express more of the logical structure of a sentence. It follows from this that with enhanced expressive power we can reduce more of mathematics to logic. This is because the greater the expressive power of a formal system, the more sophisticated our analysis of the pre-formalized concepts. For example, first-order logic (also called predicate logic) is more powerful than propositional logic.[7]

Consider the following sentence: "If Mary mucks the barn, then someone mucks the barn". In propositional logic this is represented as a conditional statement between two different propositions: P = "Mary mucks the barn"; Q = "Someone mucks the barn". The whole sentence comes out as: $P \rightarrow Q$.[8] Once we have formalized the sentence, we cannot see any connection between P and Q, although it was there in the English sentence.

In contrast, if we formalize the sentence in first-order logic, then we might assign "a" to the name "Mary", and the predicate letter "F" to "mucks the barn". Then the English sentence is represented as: $Fa \rightarrow \exists x(Fx)$. This is read, "if a has the property F, then someone has the property F". Here we can see that there is a connection between the thoughts on each side of the implication sign by the use of "F". This illustrates the fact that first-order logic has more expressive power than propositional logic. This greater expressive power entails greater powers of inference. Specifically, there will be some arguments whose validity can be demonstrated in first-order logic, but not in propositional logic. This is as a direct result of the greater expressive power of first-order logic because first-order logic can reveal interconnections that are lost in propositional logic. For example, the sentence "If Mary mucks the barn, then someone mucks the barn" is always true. However, we can only prove this by translating the sentence into first-order logic. We cannot prove this if we simply translate the sentence into propositional logic.

Frege's formal system of logic is more powerful than Aristotelian syllogistic logic. It is also more powerful than propositional logic, Boole's logic and first-order logic. Oddly, while Frege's logic is more powerful than either propositional logic or first-order logic, it was developed before either of them.[9] Frege's formal system is arguably equivalent to what we call "second-order logic" today. The major difference between first- and second-order logic is that in second-order logic we are not restricted to quantifying over variables that only pick out objects in the domain, but we may also quantify over predicates, relations and functions. For example, we may express "mucking the barn is a daily task". "Is a daily task" is predicated of the predicate "mucking the barn"; this makes it a second-order predicate.

To summarize, when Frege developed his system of logic, it had much greater expressive power than the existing Aristotelian and Boolean systems of logic. We have seen examples of what greater expressive power means. The advantage this brings is not only that we can show interconnections between sentences, and so prove the validity of more arguments, but we can also recapture more of mathematics in logic. If we have a more expressive logic, then we can reduce more of mathematics to logic. The reduction of a branch of mathematics to logic consists in a translation from the branch of mathematics into the language of the logic, together with some demonstration that we can prove the theorems of the branch of mathematics in the logic, and then translate back again to show the match. In other words, we show that the branch of mathematics adds nothing new to the logic. The mathematics can be absorbed within the logic, and so is strictly redundant with respect to the logic.

Recall some vocabulary. If we are reducing one branch of mathematics to logic, we call the branch of mathematics the "reduced" discipline. The logic, to which we are reducing the branch of mathematics, is called the "reducing" discipline. For example, if Frege wants to show that arithmetic is really logic, then he has to show that arithmetic is reducible to logic. Arithmetic is the reduced discipline, and logic is the reducing discipline.

More specifically, in order to prove that arithmetic is really logic Frege has to show that he can express concepts such as addition and multiplication in the language of logic. He also has to show that he can prove the theorems of arithmetic, in that language. For example, Frege has to show that he can prove that 2 + 2 = 4 by appeal to logic alone. If Frege can show that this, and any other theorem of arithmetic, is really a theorem of logic, then he has shown that arithmetic can be reduced to logic. Notice the switch between the notions of axiom and theorem. Axioms come at the start of the presentation of a formal system. They are the basic truths of the formal system, and if we accept them, and the rules of inference, then we accept what follows from the axioms, where "follows from" means derived by means of the rules of inference. The sentences we conclude from the system's axioms, by use of the rules of inference, are all theorems of the formal system.

In his system of logic, Frege proved the axioms of arithmetic. That is, what were previously thought of as the axioms of arithmetic are logically derived as theorems of logic; the "axioms" of arithmetic are logical theorems not arithmetical axioms. This shows that all the theorems of arithmetic are based on logical principles alone. This was an important result. Previous to Frege's work, both Peano and Dedekind had independently (around 1888)[10] developed a series of arithmetical axioms that capture the whole of arithmetic; that is, we can derive all the theorems of arithmetic from them. The axioms of arithmetic govern the notions of zero, addition, multiplication and "immediate successor" (4 is the immediate successor of 3). Zero, addition, multiplication and immediate successor are *prima facie* thought of as arithmetical concepts, not as logical concepts. To represent the axioms of arithmetic formally, we symbolize the "immediate successor" of an arbitrary number x as "Sx"; x and y vary over the natural numbers; "·" symbolizes multiplication; "$\forall$" is the universal quantifier, which we read as "for every"; "+" stands for addition; "$\geq$" means greater than or equal to; and "$>$" means strictly greater than. The Peano/Dedekind axioms of arithmetic are as follows:

Axiom 1: $\forall x(Sx > x)$
For every number x, the successor of that number is strictly greater than it.

Axiom 2: $\forall x(x \geq 0)$
Every number is either equal to 0, or is strictly greater than 0.

Axiom 3: $\forall x((0 + x) = x)$
For every number x, x added to 0 is identical to x.

Axiom 4: $\forall x \forall y((x + Sy) = (S(x + y)))$
For every x, for every y, x added to the successor of y is identical to the successor of (x added to y).

Axiom 5: $\forall x((0 \cdot x) = 0)$
For every x, multiplying 0 and x is identical to 0.

Axiom 6: $\forall x \forall y((x \cdot Sy) = ((x \cdot y) + x)))$
For every x, for every y, x multiplied by the successor of y is identical to multiplying x by y, and then adding x to that.

Axiom 7: $\forall F((F0 \,\&\, \forall x(Fx \rightarrow FSx)) \rightarrow \forall y(Fy))$
For all properties F, if 0 has the property F and for all numbers x

if x has the property F then the immediate successor of x also has the property F, then all numbers y, have the property F.

Axiom 1 prevents looping. That is, it stops the idea of the numbers increasing and either stopping, or going around in a circle. For example, a system of numbers that would defy Axiom 1 might progress 0, 1, 2, 3, 4, 5, 3, 4, 5, 3, 4, 5 .... Axiom 2 makes 0 the least number of all the numbers. Axiom 3 just says that adding 0 to a number does not change the number. Axiom 4 says that if a number x is added to the successor of a number y, then this is equal to the successor of x added to y. Axiom 5 says that 0 multiplied to any number is equal to 0. Axiom 6 says that if a number x is multiplied by the successor of a number y, then this is equal to multiplying x by y and adding x.

Axiom 7 is the most interesting. It is called the "axiom of induction". It says that if a property applies to 0, an arbitrary number and the immediate successor of that number, then it applies to all the numbers. An example of such a property, which is true of any number n, is the identity between all the predecessors of n added together and $½n(n + 1)$. This is true of any natural number n. The property is an identity statement. The axiom of induction allows us to reason over an infinite set: the natural numbers. To reason over the infinite set, we only need to look at what we call the "base case" – usually 0 – and look at an arbitrary number and its immediate successor. We then have enough information to say that the property must be true of all the numbers. This is much faster than checking all the numbers individually. This also presupposes the notion of actual infinity. The axiom of induction allows us to reason over all the natural numbers at once.

The axiom of induction also acts as a formal definition for the natural numbers in the sense that we say that the natural numbers satisfy the axiom, or model the axiom. If we are philosophically more sensitive, then we think of the natural numbers as coming before, or being conceptually more primitive than, the axiom of induction. We then say that the models that satisfy the axiom of induction do a good job of capturing our primitive notion of natural number. In fact, there is only one model, unique up to isomorphism, which satisfies all the axioms, and that is the set of natural numbers. That is, any model that satisfies the axioms will have the same cardinality, and order structure, as the natural numbers.[11] The model is unique because we cannot mathematically distinguish what we might have erroneously supposed were "different models". This is very important because Frege complains that we do not yet have, in mathematics, a good definition of number. "Number" is usually taken as a primitive term. Frege proves to us that numbers are logical entities.

A delicate issue arises here. The axiom of induction, as written above, is second-order because there is a quantifier quantifying over a predicate. Some philosophers, logicians and mathematicians object to second-order quantification.[12] We can replace the second-order axiom with a first-order axiom scheme. It looks like this:

Axiom 7': $(F0\ \&\ \forall x(Fx \rightarrow FSx)) \rightarrow \forall y(Fy)$

The *F*s in axiom 7' are then schematic letters, which we fill in with constants, that is, with defined predicates. For example, we could say, *F* stands for "is even".[13] Once we say what *F* stands for, *F* becomes a constant; its meaning does not shift. There are an infinite number of ways of fixing *F*, so there is a separate axiom for each constant *F*. Since there are an infinite number of these, we call Axiom 7' an "axiom scheme" as opposed to a single axiom.

The advantage of the axiom scheme is simply that we restrict ourselves to first-order quantification, and this gives us certain desirable (because tidy) properties of the logic. The disadvantage of only allowing the first-order axiom scheme of induction is that the set of axioms then becomes weaker, in the sense that we can construct non-standard models of arithmetic. These will be sets of numbers that are not isomorphic to the natural numbers.[14] They will have a different cardinality or a different order structure from that of the natural numbers.[15] Different progressions of "the numbers" satisfy the axioms.

It should also be noted that what is considered to be a "good" property for a mathematical system to have and what is a "bad" property varies from one mathematician to the next. For example, some mathematicians find non-standard models of arithmetic very interesting, and see the existence of such as an advantage of first-order arithmetic over second-order arithmetic. Others disagree.

Regardless of the disagreement, Frege developed second-order logic, and so was able to derive as "logical theorems" the axioms of arithmetic. Frege derived the second-order axiom of induction, not the axiom scheme of induction. Proving the axioms of arithmetic from logic is a great feat. It shows that we do not need to take these axioms as primitive, or as the ultimate basis of arithmetic. Instead, we learn that logic is the ultimate basis of arithmetic. What were thought to be axioms of arithmetic turned out to be theorems of logic. The theorems of arithmetic, such as 2 + 8 = 10, turned out to be more theorems of logic. More impressive still is that Frege proved that there were an infinite number of natural numbers. Through his proof, the notion of infinity becomes a logical notion, and arithmetic is really just a branch of logic. When Frege goes on to capture analysis, which is the study of the real numbers, he proves that there are an infinite number of these, and he voices his confidence that he can, in the future, reproduce Cantor's diagonal argument in logical notation only to show that there are more real numbers than natural numbers, thus making the infinite cardinals of Cantor all logical notions.

3. Frege's logicism: philosophical accomplishments

Frege's proof that arithmetic can be founded on logic is philosophically significant, both epistemologically and ontologically. The significance rests on the view that knowledge and justification are arranged in a hierarchy. Since logic stands at the top of the hierarchy, it is universal and perfectly general. Logic gives us the laws of thought, or the constraints on thought. By reducing arithmetic to logic, Frege shows that arithmetic is ultimately justified, since it is universal and objective. Logic, and *a fortiori* arithmetic, is objective in the sense of being based on logical objects.

What is a logical object? We normally think of "objects" as physical objects. We think of tables and chairs, what J. L. Austin calls "medium-sized dry goods". But there are other sorts of objects: abstract objects.[16] Abstract objects do not have a location in space and time. Ideas are sometimes thought of as abstract objects, and in our case numbers can be thought of as abstract objects. They are objects in three senses:

(i)   in the sense of being the referent of singular terms;
(ii) in the sense of not owing their existence to us; and
(iii) in the sense of being objects of study.

Sense (i) is motivated by how we use language, and how the grammar of our language is structured. Speaking grammatically, we say that "the number 3" is a singular term in the sentence "The number 3 comes before the number 8 in the ordinal number sequence". A singular term can be a subject, object or indirect object of a sentence. A singular term in a sentence refers to a single object. It is by virtue of grammar that we suppose it refers to a single object. We infer from this that for a sentence containing a singular term to be true, the object it refers to must exist. So we have a *prima facie* grammatical indicator that numbers are objects. Grammatically, we treat numbers as objects and grammar shapes our conceptions. We might, of course, be misled by grammar. "Unicorn" is also a singular term, but unicorns do not exist, at least according to current scientific theory. This is why grammar only provides a *prima facie* reason to think of numbers, say, as objects, and not a definitive reason. Senses (ii) and (iii) are more decisive.

Consider sense (ii). When we do arithmetical calculations we study properties of numbers, for example, we might "discover" that $387 < (567 \div 1.27) + 46$. We discover, as opposed to create, objects that exist independently of us. Arithmetical equations that are true are absolute and undeniable. As such there have to be objects that make the sentence true. These are the numbers. We refer to them to demonstrate the truth of our discoveries in arithmetic. This entails that we cannot, and should not hope to, think of alternatives to arithmetic, since the numbers exist independently of us.[17] We cannot influence or shape them. In contrast, since Euclidean geometry is not universal (there are several competing geometrical systems), we can, and do, think of alternatives.

Now consider sense (iii). What makes numbers objects of study is that mathematicians study them. When we do logic, we study the objects of logic, much as a biologist studies living things: the objects of biology. The truths of biology are discovered and tested by matching them with the objects of biology. Similarly with arithmetic; we test our arithmetic theory by checking out the theory against how the numbers work. So we study the objects of mathematics. They are independent of us, under sense (ii).

The first and second senses of "object" are interesting since the objects of logic are logical objects, and this is philosophically important. Numbers can be defined using only logical language and logical notions. Moreover, we can derive the existence of the individual numbers as theorems of logic. That is, according to the Fregean logicist, it is a logical truth, a tautology, that, say, the number 6 exists. For this reason we say, for example, that "the number 6 is a logical object"; that is, it only takes logic to show its existence and we do not need to appeal to special principles. It follows that numbers occupy a special place in our thinking. Since the objects of arithmetic are objects of logic they are universal. Logic is ubiquitous. We can help ourselves to logical inferences or principles at any time, when discussing any subject. Since arithmetic is part of this, we can count anything. Numbers are universal.

Another aspect of occupying the top of the hierarchy of knowledge is that using principles of arithmetic is never metaphorical. Let us explain this using an example. One can apply principles from mechanics to economics, and say, for example, that to every action in the market there corresponds an equal and opposite reaction. The mechanical law is that for every (physically defined) action there is an equal and opposite physical reaction. If a billiard ball hits another, then no energy is lost; the energy is simply translated into the movement of the other billiard ball, and expended in friction against the pool table and in the sound of the impact. Similarly in economics (according to some theories), we exchange money for goods. The money we pay is the value of the goods and for every item sold there is a price: an equal and opposite reaction. We are using a metaphor from mechanics and applying it to economics. In contrast, when we count, or use logic, we are not using a metaphor or simile; we directly use logic and *a fortiori* arithmetic, and this is testimony to the universality of logic, and hence logical objects. We have choices in adopting one metaphor or another for our economic theories, but we have no choice when counting. Arithmetical facts are "brute facts".

The objectivity of arithmetic truths explains why it is that we find arithmetic so compelling. It is not something we can deny. Equations such as $2 + 0 = 2$

seem undeniable and solid. According to the logicist, this is because these truths are independent of us. They depend on logical objects. We discover them, we do not invent them, so it is not in our capacity to change them by rethinking. Our grammar does not mislead us.

Our being able to give an ultimate justification for arithmetic also has implications for the epistemology of arithmetic. By giving an ultimate justification for arithmetic (i.e. by reducing arithmetic to logic), Frege showed that arithmetic is *a priori* and analytic. A truth is analytic if and only if it is true in virtue of meaning and/or logic. A celebrated example of an analytic truth is "Bachelors are unmarried men". The sentence is true by virtue of the meaning of the words in it. All we do in the sentence is unpack the word "bachelor". That is, the sentence "Bachelors are unmarried men" is an analysis of the word "bachelor". The opposite of a truth being analytic is that it is synthetic. A sentence that is synthetically true will be true by virtue of putting together independent ideas. A sentence that is synthetically true might be one that tells an empirical fact, such as "The prevailing winds come from the west". This is an observation statement. We have to observe the fact that most of the time winds blow from west to east. We put together two ideas, "prevailing wind" and "direction of wind", and we create a synthesis of the two concepts. The sentence is not true in virtue of the meaning of the terms in the sentence, so it is not analytic.

Returning to arithmetic, if arithmetic is logic, and if logic is analytic, then arithmetical truths are true by virtue of meaning. They are true by virtue of the axioms of logic, definitions and logical inference. Logical truths are trivially analytic. For example, "Either the brumish is flampy, or it is not" is an analytic truth. We do not have to have had an experience of "brumishes" or have any idea of what "flampy" means in order to divine the truth of the statement. It is a logical tautology. This time the sentence is not analytic in virtue of meaning, but in virtue of logic. When Frege derived the Peano axioms from logic, he derived them from logical axioms and definitions, using gapless proofs.

Frege called the logical axioms "basic laws" so as not to confuse them with the axioms of a particular theory. The basic laws were presented as logical in the sense of universal, independent and analytic. A basic law is absolutely primitive. In contrast an axiom is an axiom of a formal system. An axiom is not supposed to be absolutely primitive, only primitive *relative* to the theorems of the formal system. To ask about theorems, we ask about the axioms of the system and the rules of inference. To ask about the axioms, we have to step outside the system, usually to another formal system. To ask about basic laws we engage in philosophical discussion, because we have to step outside all of mathematics and logic.

Apart from basic laws, Frege allows definitions in his formal system. In Frege's formal system, definitions were only shortcuts, so were ultimately dispensable. For example, the definition of "bachelor" is "unmarried man". If we like, we could dispense with the word "bachelor" in all our writing, and replace it with "unmarried man" without losing the meaning or truth of sentences. Frege's formal system of proof is "gapless". Each move in a proof is accounted for by appeal to a previously accepted rule of inference. In fact, in Frege's formal system, there is only one rule of inference, and this is *modus ponens*. The rule *modus ponens* says that if you have a conditional and, independently, the antecedent of the conditional, then you may infer the consequent of the conditional.[18] The gapless proofs ensured against quick reasoning, which might involve appeal to some unexamined principle, which might turn out to be adding some new concept to the system, which would make the derived "truth" synthetic. This was a crucial part of the demonstration that the truths of arithmetic are analytic.

Since analyticity and syntheticity are exclusive of one another, another way to demonstrate that arithmetic is analytic is to show that it is not synthetic. Arithmetic could be synthetic for two different reasons: arithmetic could be an empirical science or arithmetic could depend on spatiotemporal intuition. We shall start with the first. One claim that the logicist has to justify is that arithmetic is not empirical. This is easily done. The logicist's hierarchical conception of knowledge and justification implies that arithmetic is not *a posteriori*, since establishing, or discovering, the truths of arithmetic does not rely on sense experience,[19] and if a truth is known *a posteriori* this precludes it from being analytic. A truth is known *a posteriori* when we have to use sense experience to know it. This is not to say that we cannot learn some arithmetic using "empirical experimentation". Rather, the point being made by the logicist is that empirical experiments, which necessarily rely on sense experience, will not get us very far in our understanding of arithmetic. For example, we might try to teach a child that 8 + 8 = 16, by having the child count out eight marbles, and another eight marbles, putting them together and then counting, and discovering that there are 16 marbles. We might have the child repeat the experiment for several types of object, until the child is convinced, empirically, that 8 things plus 8 other things equals 16 things altogether. The problem with this method is that the child does not *necessarily* learn general principles about adding. The child is only empirically justified in thinking that adding 8 objects with another 8 objects will make 16 objects. Empirically speaking, a new experiment has to be designed for adding 8 objects to 9 objects. The experimental method only gets us so far.

To learn general principles about addition, one needs to abstract from our experience and generate general principles such as axioms, and these are analytic. They go well beyond experience, and are not testable by physical experiment or physical observation, especially in the case of infinite numbers. The laws of logic, and *a fortiori* the theorems of arithmetic, which used to be thought of as axioms, cannot be inducted, or deduced, from physical experiments, according to the logicist. In fact, to think one can is just wrong-headed, again according to the logicist. Instead, we come up with the general principles, and it is these that give the meaning of numbers, adding and so on.

Under the logicist view, not only does arithmetic not ultimately depend on particular empirical experiments, but arithmetic does not rely on spatiotemporal intuition in Immanuel Kant's sense either. This would be the second way in which a body of truths could be synthetic. It is much harder to show that arithmetic is not synthetic in this way. Frege takes issue with Kant's belief that both arithmetic and geometry are synthetic. Truths are synthetic (which is the opposite of analytic) if they ultimately depend on either sense experience or on Kantian intuition. We say "Kantian intuition" because Kant uses the term "intuition" in a rather special way. For Kant, arithmetic depends on what he called temporal and spatial intuition. In Kant's technical sense, "intuition" is the bridge between sense experience and pure reasoning. Intuition makes it possible for us to apply our reasoning to the physical world around us. By "intuition", he means something that is available to everyone in the same way; that is, for Kant, we all have the same spatial and temporal intuition. We share these intuitions exactly, so they are universal, in the sense of "common to us all". Moreover, this is a necessary fact about how we represent the world to ourselves. What temporal intuition gives us is a sense of ordering: of one moment coming before the next in time. This is analogous to the numbers: one is less than another. The spatial intuition gives us a sense of one number being "located before" another on the number line. Interpreting Kant on his notions of temporal and spatial intuition is subtle work.[20] Frege did not think that something as basic as arithmetic ought to depend on intuitions, either in Kant's sense or in any other looser sense.[21] For Frege, arithmetic is simply part of basic logic, not requiring intuition at all, because of the point made above about not being able to think of alternatives to arithmetic. This supports the idea that arithmetic is logic and not a special science relying on intuition. We have to be careful. Frege was not entirely rejecting Kant's ideas about intuition;[22] rather, he thought that Kant had underestimated the power of logic. When Kant was writing, the study of "logic" was more or less confined to Aristotelian syllogistic logic. Kant's ideas about arithmetic requiring spatiotemporal intuition did not come as a result of underestimating the power of Aristotelian syllogistic reasoning; rather, Aristotelian syllogistic reasoning does not formally represent all there is to logic. It is second-order logic, as presented in Frege's formal logical system, that fully represents "logic", and arithmetic is reducible to this logic.

If Kant had been introduced to Frege's formal system, would he have sided with Frege or not? One possibility is that Kant would have sided with Frege and said "Ah, you are right! I did not have at my disposal a full logical system.

I only had Aristotelian syllogistic reasoning. Now I see that arithmetic does not rely on intuition, but is analytic". In this case, Kant would have recognized that thinking that Aristotelian logic exhausted "logic" had been a mistake; "logic" has much wider compass. Alternatively, at our imaginary meeting, Kant could have denied that Frege's formal system represented "logic" at all. Kant might have stuck to the position that Aristotelian syllogistic logic is all there is to "logic"; the rest is mathematics. If this were Kant's reaction, then, when presented with Frege's formal system, Kant would have said that what Frege calls "logic" is mathematics, and that to grasp it one needs spatial and temporal intuition. In this latter case, the debate is over what counts as "logic" and what "analyticity" is, rather than about the scope and formal representation of "logic". We shall never know how Kant would have reacted.

Either way, Frege is explicitly taking issue with Kant. This is a bold move. Kant's philosophical remarks about mathematics were considered to be a very important reference point for philosophers. According to Alberto Coffa, for "better or worse, almost every philosophical development since 1800 has been a response to Kant" (1991: 7).[23]

To sum up, the philosophical significance of logicism accounts for its appeal. Logicism gives some account of our reluctance to question arithmetic. Unfortunately, there are some deep problems with logicism.

4. Problems with Frege's logicism

The most devastating problem with Frege's attempt to prove that arithmetic is really logic was discovered by Russell, who had a manuscript version of Frege's *Grundgesetze*. Before the second volume of *Grundgesetze* was published, Russell wrote to Frege explaining that one could derive a contradiction in Frege's formal system. This has come to be known as "Russell's paradox".[24] Since Frege received the letter too late to make major changes to the second volume, he acknowledges Russell's discovery in an appendix to that volume of *Grundgesetze*. Frege immediately recognized the gravity of Russell's discovery:

> Hardly anything more unfortunate can befall a scientific writer than to have one of the foundations of his edifice shaken after the work is finished.
> This is the position I was placed in by a letter of Mr. Bertrand Russell, just when the printing of this volume was nearing its completion.
> (Frege 1952: 214)[25]

The problem has to do with Frege's basic law V, concerning extensions of concepts. Frege's basic law V is:

$$\forall F \forall G((\text{Ext}F = \text{Ext}G) \leftrightarrow \forall x(Fx \Leftrightarrow Gx))$$

F and G are schematic letters standing in for concepts; that is, they can be replaced by predicates, relations or functions. Basic law V says that for any concepts F and G, the extension of F is identical to the extension of G if and only if all the objects falling under the concept F are the same as all the objects falling under the concept G. In other words, basic law V just explicates the notion of "extension of a concept". It seems innocuous enough. It seems to be trivially and obviously true.

Explaining further, the notion of extension is contrasted to that of intension. Both extension and intension have to do with how the concept is presented to us. The extensional presentation of a concept is just a list of the objects falling under the concept. In contrast, an intensional presentation of a concept gives a characterization of the concept, which then allows us to pick out which objects fall under it. For example, someone might give a list of five people who are invited to dinner. The list is the extension of the intensionally given concept "person invited to dinner".[26] We can plug this example into basic law V. Call "F" the concept "person invited to dinner". Call "G" the concept "member of the poetry club". The list of people invited to dinner is identical to the list of people in the poetry club if and only if the two lists have the same people picked out. When this happens we say that, logically, "the concepts are equivalent", where equivalent means "similar in some, possibly many, respect(s)". In this case, the concepts are identical in their extensions, so the concepts are logically interchangeable (we do not care which way you call them, or what means you have of picking them out). That is, how one comes to generate the list – by looking at the list of members of the poetry club, or looking at the guest list for the dinner party – is not relevant to the extensions of the intensional characterization. We just want to be able to generate the list. Basic law V is intended as a law of logic. It says that it is a matter of logic that how we express ourselves (how we intensionally represent concepts) is irrelevant, provided we pick out the objects we want. Two concepts are identical in extension, that is, logically indistinguishable, if they pick out the same objects. Logic is extensional: blind to intensional presentation. The concept 2 + 2 is logically equivalent to concept 4. The concepts have identical extensions, since "2 + 2" picks out the same object as "4" does.

Basic law V might seem trivial and obvious. This is partly because nearly all mathematicians presuppose that their systems are extensional systems, so they endorse, implicitly or explicitly, something like basic law V. Frege recognized this. It was for this reason that Frege thought that basic law V could be accepted as a logical principle. Frege just made the principle explicit. Basic law V, and other axioms like it, are sometimes called "naive" (comprehension)[27] principles. The principles are naive because they seem quite obvious but lead to contradiction.

From basic law V we can generate a contradiction because there is no restriction on the sorts of concept we are allowed to countenance in Frege's formal system.[28] *Any* two concepts have the feature of being identical in extension (logically indistinguishable) just in case they pick out the same objects. We can substitute what we like for F or G provided we can express, or formulate, the concept in the language. Recall the universal quantifiers at the beginning of basic law V, $\forall F \forall G$: "for any concepts F, for any concepts G". Also recall that in classical logic we can always derive an existential sentence from a universal: $\forall x(Hx) \vdash \exists x(Hx)$ is logically valid ($\vdash$ is read as "we can syntactically (following logical rules of derivation) derive"). The expression $\forall x(Hx) \vdash \exists x(Hx)$ is read: "All objects in the domain have the property H, therefore, there is something in the domain which has the property H". In basic law V, the universal quantifiers quantify over second-order objects, that is, over concepts. So, the universals in basic law V imply, with a little manipulation, the existence of any concepts we think of out of the blue to substitute for F and G. Basic law V licenses us to think up any concept we can express in the formal language, and guarantees the existence of an extension for that concept. Note that the extension might be empty, in case there is nothing to pick out. A contradictory concept picks out the empty set. For example, the concept of "things not identical to themselves" picks out the empty set.

We can now turn to Russell's objection to basic law V. Consider the concept "the set of all the things not in their own extension". Most sets do not have themselves in their own extension. For example, the set of members of the poetry club is not itself a member of the poetry club. Therefore, the set is not in its own extension. In contrast, we can think of concepts that do have themselves in their own extension. An example would be the concept "is an infinite set". The set of "infinite sets", is itself infinite, and therefore is in its own extension. Returning to the notion of "sets that do not have themselves in their own extension", we can collect *all* of these together under one concept: "all sets not in their own extension". This forms a set. Call it "R", for Russell. We now ask the question whether R is in its own extension. If R is in its own extension, then it ought not to be by virtue of the meaning of the concept. If R is not in its own extension, then it ought to be by virtue of what is included under the concept. We have a contradiction.

Basic law V is not as innocent or obvious as we thought. Part of the problem is that it allows any concept at all: it is unqualified. Basic law V does not say "all concepts except for ...". Adding such a clause to basic law V is not an option. If it had exceptions, then it would not be a candidate for being a principle of logic, for basic laws are supposed to be entirely general, and a purported "law of logic" that has some exceptions is not general or universal.

Frege tried to repair the damage by suggesting an alternative to basic law V, from which Russell had derived a paradox. Unfortunately, the alternative also turned out to be contradictory. Frege was in deep despair about his project and he did not try to repair it further. In fact, he did not publish again for fourteen years. When he finally did, he tried to re-found arithmetic in Euclidean geometry, rather than in logic. This is because Frege was still convinced that the mathematical analysis of the notion of number was inadequate. The analysis of the notion of number had to be given by appeal to something more basic than the notion of number itself. He decided that Euclidean geometry must be more basic, in terms of giving a justification for our number concepts. His further project was neither fully developed nor seriously taken up by any followers, so it was not pursued.

Apart from inconsistency, there is a second major problem in Frege's presentation of his philosophical view. It is quickly stated. In the literature, it is referred to as "the Julius Caesar problem". The problem is this. From within logic itself, as we are sitting at the top part of our hierarchy of knowledge, we cannot tell if an arbitrary object that is presented to us, such as Julius Caesar, is a number or not. This sounds very odd. But the point is that *logic by itself* cannot tell us really what sorts of objects numbers are; at least not enough to distinguish them from other sorts of objects. If we rely on common sense then we know that Julius Caesar is not a number, but logic alone cannot tell us that.

This marks a failure in Frege's analysis of the concept "is a number". This is because we have to go down the hierarchy of knowledge in order to tell us something about the upper levels of the hierarchy. So logic is not a self-sufficient discipline; it requires the help of the lower levels. This is a problem for Frege, because one of the motivations for Frege in developing logicism is that there is no adequate theory of number, and it turns out that, by his own lights, his theory of number is not adequate either.

Frege was well aware of this problem. He discusses it in *Grundlagen* §56, and thinks he has solved the problem by §66. However, the way in which he solved the problem is to use the notion of the "extension of a concept". Frege then had to introduce the notion of extension of a concept to his logic. To do this he presented basic law V, and we know where that led.[29]

5. Whitehead and Russell's logicism

Whitehead and Russell[30] picked up where Frege had left off. They decided to develop a logical system that was more elaborate in two respects. Their ambitions were greater than those of Frege: they sought to reduce all of mathematics to their formal system. For this they needed great expressive power in their logical language.[31] The other consideration which makes for a more elaborate formal system than Frege's is that Whitehead and Russell were adamant that they should avoid paradox, so there are no naive principles. Unfortunately, the principles that replace naive comprehension are intuitively less obvious and therefore arguably not universal, and thus not logical.[32]

The formal system developed by Whitehead and Russell is called "type theory". Type theories, of one sort or another, are used today by computer scientists. Whitehead and Russell developed two type theories: the simple type theory and the ramified type theory. The ramified type theory is more elaborate, as the name suggests, and was developed to allow a more fine-grained analysis, and organization, of mathematical concepts. We shall not discuss the finer details of type theory here since these will not be important to the overall philosophical position, but we do need to get some feel for type theory. In type theory one is entirely explicit about what type of thing falls under a given symbol. For example, it was mentioned above when discussing Frege's logic, that "F" and "G" are used by Frege to include predicates, relations and functions. In type theory these things are kept quite separate, for they are of different types. We do this in any formal system implicitly by using different fonts, symbols and alphabets. In type theory, we are very explicit about rules of inference and axioms governing different types. Moreover, in order to avoid paradox there is a strict level system where predicates are only allowed to predicate over things at a lower level. This is the secret to avoiding paradox. We are not allowed, by the "grammar rules" of the type theory, to ask of a set at one level whether it belongs to itself (at that same level), so we cannot get started on the reasoning that yields paradox. Whitehead and Russell were ingenious in designing the grammar (the notion of a well-formed formula) in such a way that concepts that lead to paradox turn out to be ungrammatical, and therefore are considered to be nonsense. The rules for making a well-formed formula, that is, the rules of grammar for the formal language, concern the types of symbols used to form expressions.

As a result of this grammatical, type-theoretic, elaboration, axioms have to be given to govern what happens to things of different types, and at different levels. For example, separate axioms have to be given about the extensions of functions and the extensions of relations. Before explaining Whitehead and Russell's type theory explicitly, notice that the vocabulary has shifted. We have moved from Frege's basic laws to Whitehead and Russell's axioms. Both formal systems are axiomatic, in the sense of explicitly listing basic principles and one or more rules of inference, and then being able to deduce the theorems from those.[33] The difference between axioms and basic laws is philosophical. Basic laws are laws of logic and, according to the logicist, logic plays a special role in our hierarchy of knowledge. There is nothing more basic than logic, if we think that logic is what gives us minimal constraints in reasoning.

LOGICISM   67

In contrast, if we are discussing an arbitrary formal system that is presented axiomatically, then the axioms are the fundamental assumptions of that formal system. So the axioms are basic *relative to* a formal system, as opposed to Frege's basic laws, which are fundamental to all systems. Because Whitehead and Russell's type theory was clearly different from Frege's system, we avoid begging any questions by calling the basic principles of the system "axioms" rather than "basic laws", and then enquire separately whether they deserve the philosophical status of Frege's basic laws.

Let us turn to the semantics of Whitehead and Russell's type theory. We shall not be using Whitehead and Russell's notation because this is not used much presently, except by Russell scholars. Also be warned that we are looking at the simple type theory, and omitting many details in the hope of giving an understandable first impression. Further details can be added on when studying Whitehead and Russell in depth.

We number each type. Individual objects are assigned type 0. Physical objects, people or abstract objects can be of this type. A relation between two objects of type 0, such as "is the mother of", is of type <0, 0>. We use the angled brackets to show that the order of the members is to be respected. In the case of the relation of motherhood, the order is important. There is a difference in truth-value between saying that Elizabeth is the mother of Bertrand, and that Bertrand is the mother of Elizabeth. A three-place relation between objects of type 0, such as "lies between", is of type <0, 0, 0>. We want to talk about objects, so we predicate properties of them. For example, we might say that the chair (which is of type 0) is red. Red is of type (0): it predicates over objects of type 0. A predicate of predicates, such as "is a colour", is of type ((0)). A relation between two predicates of type (0) is of type <(0), (0)>. An example of such a predicate is "is a darker colour than". In principle, we can extend these by adding brackets, and by adding more things of type 0. We might think of this as expanding upwards and sideways, respectively. For example, we extend the type <(0), (0)> upwards by adding brackets. We extend the same type sideways by adding more objects: <(0), (0), (0)>.

...	...	...	...
(((0)))	<(((0))), (((0)))>	<(((0))), (((0))), (((0)))>	...
((0))	<((0)), ((0))>	<((0)), ((0)), ((0))>	...
(0)	<(0), (0)>	<(0), (0), (0)>	...
0	<0, 0>	<0, 0, 0>	...

We can also have "mixed" types, for example if we want to say that red (of type (0)) is predicated of individual object (of type 0), then this is a relation between things of different type. This relation is of type <(0), 0>. It is a relation between a predicate and an object. This type assignment system is important because we can make rules for the expansion, and then we can prevent paradox. We prevent paradox by forbidding certain sorts of type. We do this by stipulating axioms and grammar rules about which types are allowed, and which are not. In the Whitehead and Russell type theory we may not discuss the notion of "all properties", since it has no type; we would not be able to guess how many brackets to put around the original 0.[34] If we thought we had guessed correctly, then there would be a question about whether that new type itself were a property. Moreover, we could generate a paradox by asking if the property of "not being a property" is itself a property. All of these dangerous notions are forbidden through the rules for piece-by-piece construction of expressions of different types. It is because we can order the types that we can build piece-by-piece; details are not important here. Once we have an ordering on the types, there are strict rules about which types are allowed, given a certain type, the idea being that some types are ordered higher than others, and a type can only be built up from types of lower order. Moreover, a type can only have in its "range of significance" certain lower-order types. "Range of significance" is Whitehead and Russell's expression. It just means that an expression of order-type 3, say, can only apply to expressions of order-type 2, 1 or 0. In our notation, an expression Fa of type $(((0)))$ can only apply to as of type $((0))$. This notion of "range of significance" is particularly important for the universal quantifier.[35] The quantifiers are assigned a type, just as any other symbol is. In fact there are many types of quantifier. To make a well-formed expression, the quantifier has to be of type higher than the variable that follows it. This is what is missing in Frege. For example, in type theory $\forall x$ is allowed provided x is of type lower than $\forall$. Now x could range over objects, so be of type 0. Or x could be a predicate, so of type (0). In the first case, $\forall$ has only object-level variables in its range. In the second case, $\forall$ has predicates in its range. Because of these higher-order quantifiers, the language of type theory has great expressive power. This is what gives the type theory the power to absorb most of mathematics.

Whitehead and Russell's type theory is a formal theory that is so expressive as to be foundational to most of mathematics: we can do most of mathematics in type theory. Several modifications have been introduced more recently, and various sorts of type theory are used by computer scientists today. However, there is some philosophical unease about the type theory acting as a foundation to mathematics in the special sense that a logicist wants. Recall that in the case of Whitehead and Russell, the logicist claim is that all, or most, of mathematics is really logic.

6. Philosophically, what is wrong with Whitehead and Russell's type theory?

As we saw, type theory is very powerful: powerful enough to be considered to be a foundation to mathematics. It is probably consistent.[36] It is even a useful theory. The development of type theory was supposed to prove the philosophical claim that mathematics is essentially logic. To prove that mathematics is essentially logic we need three things: (i) we need a reduction of mathematics to a founding discipline; (ii) the founding discipline must be consistent; (iii) we need support for the claim that the founding discipline is logic, in the philosophical sense of "logic". Whitehead and Russell's type theory fairly comfortably meet the first two requirements but is criticized by philosophers for not really being logic. The complaints are made on two counts. One is that Whitehead and Russell could not prove that the natural numbers formed an infinite set, which Frege was able to do in his logic. In fact, Whitehead and Russell were unable to prove that there exists any infinite set at all; they had to add an axiom that states explicitly that there is an infinite set.

The philosophical criticism says that it is not *prima facie* a logical truth that there exists an infinite set; it is only a mathematical truth. It is not the business of logic simply to declare which sets exist and which do not. If Whitehead and Russell were to really succeed in their philosophical aims, then they would have to prove that there is an infinite set from logical principles alone. They cannot assume the existence of an infinite set as a matter of logic. On the other hand, it is indispensable to a lot of mathematics that there should be an infinite set. Thus, any reducing discipline that posits an infinite set by virtue of the axioms of the system is more properly described as a mathematical system as opposed to a logical system.

The second complaint against (iii) is similar, but more technical. Critics of the Whitehead and Russell project question the philosophical status of the axiom of reducibility. The axiom was proposed in order to overcome the following difficulty: while we can derive the natural numbers at various levels in the hierarchy, it is not clear that they are the same ones each time. This is because the types can talk about things of lower type, but we cannot talk about sub-types being *the same* at all levels, because we cannot talk about *all levels* without the axiom of reducibility. The axiom of reducibility guarantees that we have exact copies of the numbers at every level. The critics of Whitehead and Russell's philosophical project point out that this too is not obviously a matter for logic to decide. It looks more like a convenience, or a mathematical fact, at best.[37]

Russell was aware of these criticisms and his final response was to admit that this was a problem. He distinguishes logically necessary axioms from what he calls "empirical axioms" in the following way. Logically necessary axioms are much like Frege's basic laws. For Russell, they are indubitably logical laws. In contrast, empirical axioms do not enjoy the same philosophical status as logically necessary axioms. That is, there is some lingering doubt as to whether they are logical or properly mathematical (Potter 2000: 160). They are justified empirically in the sense of being quite useful in retrospect. Ultimately, there is a problem with this attempt at logicism. Mathematics is reducible to a formal system that includes both necessary axioms and empirical axioms. Moreover, the empirical axioms are ineliminable. Therefore, the Whitehead and Russell attempt at proving logicism fails because all it shows is that mathematics is reducible to type theory, and type theory is just another mathematical discipline.

7. Other attempts at logicism

In 1983 Crispin Wright suggested a possible repair to Frege's project.[38] Wright suggested that we make a change to Frege's formal system by removing basic law V, and replacing it with Frege's "numbers principle". This new formal system has the technical merit of being consistent, and from this set of axioms (Frege's first four basic laws plus the numbers principle) we can derive the Peano axioms as theorems of the formal system. For a presentation of Frege's formal system of logic see his *Begriffsschrift* (1976) and *Grundgesezte* (1980b).

There are a number of interesting things to say about this brilliant suggestion. Technically, it works. That is, we can indeed derive the Peano axioms as theorems of a more "primitive" formal system.[39] However, we now come up against the same complaint as we saw made against Whitehead and Russell's type theory. The question is whether this more primitive system is really logic. To answer this, we have to look closely at the numbers principle.

The numbers principle is:

$$\forall F \forall G((NF = NG) \leftrightarrow (F \approx G))$$

This is read "for all concepts F, and for all concepts G, the number of Fs is identical to the number of Gs if and only if F and G can be placed into one-to-one correspondence". This recaptures Cantor's notion of the size of a set. Rather than discuss size as an absolute cardinal notion, it compares the sizes of two sets to each other. The sets are picked out by the concepts F and G. So with the numbers principle we have the notion of two sets being of the same size. For example, take the concept F to be "guest coming to dinner"; the concept G might then be "is a place set at the table". Then we would say that the number of guests is identical to the number of places set if and only if every guest has one and only one place set for him, or her, at the table. There are not more places set than there are guests, and all guests have a place. The question for the neo-Fregean, as Bob Hale and Wright call themselves,[40] is whether the numbers principle is a basic law. That is, Hale and Wright want to justify the claim that the numbers principle is a law of logic.

In terms of loyalty to Frege's original intentions, the numbers principle is significant. Frege had proved this principle in his system, and used it to derive the axiom of induction in Peano arithmetic. So it was already a theorem in Frege's formal system. Furthermore, Frege discusses the numbers principle in *Grundlagen* (1980a). In Frege's discussion of the principle in *Grundlagen*, he says that while it is very obviously true he is not convinced that it is obviously a logical principle, as opposed to a principle of arithmetic (which he has not yet proved is simply logic, and he does not want to make any assumptions). This is because the numbers principle mentions the notion of number, and this has not yet been defined as a logical notion. To prove that the numbers principle is a principle of logic, he later derived it from basic law V in *Grundgesetze*. This derivation proved not only the obvious truth of the numbers principle, but also its logical pedigree. In light of the contradiction derived from basic law V, the *logical* pedigree of the numbers principle is again in question.

The notion of definition deserves attention, for we might think that the numbers principle is a type of definition. If it is, then we need not defend it as a basic law, since Frege says that the truths of logic are derivable from basic laws and definitions, using his gapless proof system, which is very rigorous. Unfortunately, this will not work. For Frege, a definition should be strictly redundant, and it should allow us to "individuate" the objects being defined; that is, we should be able to pick out, or recognize, the objects. To do this a definition has to tell us when something that is presented falls under the definition or not, and the definition should tell us when what we thought were two separate objects are really the same object. The numbers principle does the latter, but not the former. Furthermore, there is no other principle that qualifies as a logical principle, or proper definition, in the offing that tells us when something is a number, as opposed to some other sort of object. This is another way of thinking of the Julius Caesar problem. Hale and Wright have possibly finally put the Caesar problem to rest in "To Bury Caesar ..." (2001a). However, we should show caution, for among philosophers there is always room for further debate.

But this is not enough. Even if we solve the Caesar problem, this will make the numbers principle, together with some other considerations, a means of individuating objects, but it will not assure us that numbers, "defined" via the numbers principle, are logical objects. Remember that the numbers principle is being put forward as a basic law, not as a mere redundant definition. We still have to deal with the philosophical status of the numbers principle. If the principle is a logical principle, then, according to Frege and the neo-Fregeans, it has to be *a priori* and analytic. The numbers principle is fairly uncontroversially *a priori* in the sense that knowing that it is true, or even just understanding it, does not require sense experience. It is not something that is confirmed or denied by the physical world around us, which we apprehend with our senses. More crassly, we could not design a laboratory experiment to prove or disprove the numbers principle.

Analyticity is trickier. Frege said that an idea is analytic if it can be proved from the basic laws of his formal system, since each of these is clearly analytic, and the proof system does not allow any presuppositions aside from the laws and redundant definitions. Every theorem derived from the basic laws and definitions is derived using only *modus ponens*. This characterization of analyticity does not help in this case because we are proposing a *new basic law* that is not provable from the other laws.

Another characterization Frege gave for analyticity is that an idea is analytic just in case it is not synthetic. That is, we can understand it without having had any particular sense experience of the world, and without any intuition (each of which is sufficient to make an idea synthetic). Since the numbers principle is *a priori*, the dispute about the status of the numbers principle really has to do with this notion of intuition. Elaborating on the earlier discussion concerning Kantian intuition, Frege uses the word "intuition" in two different ways (Goethe 2001): the first is as a sort of insight or feeling and the second is Kant's technical notion of temporal and spatial intuition. Let us see if we can rule out "intuition" interpreted as "feel" when we talk about understanding the numbers principle. We ask if it takes insight, or some feel for what number are, to recognize the numbers principle as true. Accepting the numbers principle as true does not depend on feel, since feelings or gut instincts are notoriously unreliable. Different people have different feelings about what they guess is true or false. Mathematical truths are objective, and therefore cannot depend on intuition in the sense of feelings. In fact, to argue that the numbers principle is true by appeal to gut feelings or intuition is to reverse the order of justification. We really want to know if we can acquaint ourselves with the concept of number by studying the numbers principle, without a prior conception of number. The answer to this is less obvious, and requires that we show that the principle is analytic. Analytic truths are sentences that are true just in virtue of meaning. Straight definitions are analytic. However, we saw above that we cannot argue that the numbers principle is a definition.

Wright gives an indirect argument to the effect that the principle is analytic. The argument is indirect in its structure. He argues that the principle cannot be synthetic, and is therefore, analytic.[41] That is, he argues that accepting the numbers principle is not a matter of appealing to Kantian spatiotemporal intuition. Instead, the principle is a special sort of definition. He calls it a "context principle". The argument concerns the structure of the principle. Let us look at it again: $\forall F \forall G((NF = NG) \leftrightarrow (F \approx G))$. In terms of structure this is not a straight definition because the biconditional (symbolized "≡" in some texts and "↔" in others, and read "if and only if") is not the main operator in the sentence. Straight definitions take the form ... ↔ ..., where the biconditional is the main operator. Instead, the pair of universal quantifiers are the main operators. The term being defined, the "definiens", is on the left and how we are to understand the term, the "definiendum", is on the right. In the numbers principle, the "definition" is couched within the scope of two universal quantifiers, which give context to the definition. Examining the principle more closely, and looking inside the scope of the quantifiers, that is, just at the expression "$(NF = NG) \leftrightarrow (F \approx G)$", we have a candidate for a definition. It is still not quite a straight definition because the same letters appear on both sides of the biconditional, so the "definition" looks circular. Wright argues the definition is not circular because if we separate out the two conditions of the biconditional the resulting parts make two very different claims. The direction "$(NF = NG) \rightarrow (F \approx G)$" is an epistemological claim. If we have identical numbers belonging to the concepts F and G, then we can place the objects falling under those concepts into one-to-one correspondence. This is how we know, and justify, that the numbers of Fs is identical to the number of Gs. The reverse direction "$(NF = NG) \leftarrow (F \approx G)$" is an ontological claim: if two concepts can be placed into one-to-one correspondence, this tells us that the cardinal numbers of the two concepts are identical. The cardinal numbers are objects because, among other things,[42] we say "*the* number of Fs". That is, for reasons of grammar (being formally represented by NF) "the number" refers to an object. This direction of the arrow makes the principle look like a definition. The other direction is what justifies the claim that the definition is analytic. The notions of identity and one-to-one correspondence are not based on spatiotemporal intuition. They are pure analytic concepts, or so the argument goes. The argument rests on an analysis of the structure of the principle.

Unfortunately, this approach faces the following "bad company" objection. The "bad company" objection is most obviously made by George Boolos in "Is Hume's Principle Analytic?" (1998a).[43] Boolos calls the numbers principle, "Hume's principle" because Frege acknowledges that the suggestion for the numbers principle can be traced back to Hume. Boolos's objection is that there is nothing to suggest that Hume's principle is analytic, except for prior acquaintance with the notion of cardinal number. This is because there are several principles that we can suggest, which share the structure of Hume's principle (alias the numbers principle), but that contradict it. So the structure is not enough to single out Hume's principle over other competing principles that share the structure. The good base of second-order logic (Frege's first four basic laws) is fine; we can accept this as generating only analytic truths. However, if we want to add to it, by attaching a principle with a certain structure, then we have a lot of opposing, mutually exclusive candidates.

Boolos calls principles that share the structure of Hume's principle "abstraction principles". By adopting this vocabulary, he wants to distance himself from Wright's claims about contextual definitions. The relevant structure is that the principle should have two universal quantifiers, which are the main operators. Within the scope of those quantifiers we have a biconditional as the main logical connective. On one side of the biconditional we have an identity between two things; on the other side we have an equivalence relation. Recall that a relation is an "equivalence relation" just in case there are some respects in which the two sides are similar. That is, the principle tells us that these respects are strong enough to make identity between the "two" objects, demonstrating when "they" are really one object. Abstraction principles are a way of telling us that we do not care about any other differences. For example, we might say that the number 3 referred to by the representation **3** and the one referred to by the representation *3* are the same 3. We do not care about which font we use to represent 3. The 3 is the same 3 no matter how many times we type 3, or where we type 3, or whether we symbolize 3 in some new way, for example, 1 + 1 + 1. The numbers principle tells us that two numbers attending F and G are identical just in case they can be put into one-to-one correspondence.

An abstraction principle that shares the same structure as the numbers principle is basic law V. But Wright rules this out since it leads to contradiction. Basic law V cannot be true, therefore it cannot be analytic. However, there are abstraction principles that can be added to the good base of second-order logic that make a consistent formal system, but are inconsistent with the numbers principle (or Hume's principle). Here is an example, which the reader need not be able to follow. The philosophical lessons can be made clear without the details. Consider the parities principle: the parity of F is identical to the parity of G if and only if F and G differ evenly (i.e. one minus the other results in an even number). In symbols we can write $\forall F \forall G((PF = PG) \leftrightarrow E<F, G>)$: "[T]he concepts F and G *differ evenly* if the number of objects falling under F but not under G or under G but not F is even (and finite)" (Boolos 1998c: 214–15). The parities principle has two universal quantifiers as main operators. Within the scope of the quantifiers we have an expression where the biconditional is the main connective. "E" (read "differ evenly") is an equivalence relation. All even numbers are equivalent with respect to the property of being divisible by two and resulting in a whole number. On the left-hand side of the biconditional we have an identity. So, we have the correct structure. If we add the parities principle to the good base of second-order logic, then we have a consistent formal system. However, the parities principle is rather strange intuitively. It says that all pairs of numbers that have the same parity are identical: they are exactly the same, indistinguishable. For example, the parity of 8 and the parity of 4 are identical, the parity of 7 and the parity of 3 are identical. The parity of 35 and the parity of 7 are identical. This makes us uneasy. This is because the parities principle is inconsistent with Hume's principle (the numbers principle) because one of the theorems we can derive from it, together with the good base of second-order logic, is that there are only a finite number of numbers (Boolos 1998b: 215). Infinite numbers do not have parities. Recall that one of the technical merits of logicism is that we can derive the infinity of the natural numbers from the numbers principle. Both principles are consistent with Frege's first four laws, so we cannot rule out the parities principle on the grounds that it is contradictory and therefore false. On the merits of structure alone, the numbers principle keeps bad company.

The lesson Boolos, Wright and Hale agree to draw from this is that structure and consistency alone are not enough to guarantee that the numbers principle is analytic; there has to be something else. "Obviousness" will not do, since this is a psychological description, and what is obvious to some people is not obvious to others. The whole point of logicism is to sanction our feeling of obviousness with regard to the numbers principle. Rather than pursue Wright and Hale's response to Boolos's bad company objection, let us turn to another response.

Köhler (whom we met in Chapter 2), suggests an ingenious solution, answering to concerns about both realism and logicism. Recall that Köhler (2000) says that there is a "rational intuition" that is required in order to recognize the truth of a logical principle. Moreover, the "rational intuition" does not offend against analyticity. That is, anything that is true according to rational intuition is analytic. Anything that is true for other reasons, requiring spatio-temporal intuition, for example, or observation, is synthetic. One can appreciate the point. It seems that some people simply have mathematical insight, or intuition; others lack this faculty. Those who lack it are those who simply did not do well in mathematics classes. Those who have the intuition see the mathematical structures, and they are very immediate to them. Obviously, they do not see the mathematical structures with their eyes, but they see them in the mind's eye, through, it is argued, the faculty of rational intuition.

There are a number of questions we could raise about this intuition, such as why some people have it and others do not, what it takes to develop it and so on. These are largely empirical and/or psychological, and possibly answered by genetic science and evolutionary science. What is more important to us here is whether positing such a faculty endorses logicism. It will speak for logicism just in case we are convinced that the truths apprehended by this faculty are still analytic. Here is the argument. We could say that any idea that can be apprehended using reasoning alone is thereby analytic, for it seems that this is what "analytic" means in this context. A truth is analytic just in case it is true in virtue of the analysis of concepts. This is contrasted to synthetic truths, which require some sort of insight having to do with our orientation in the world, and our ability to navigate in this physical world. So the numbers principle seems to be analytic in the right way. The numbers principle analyses our conception of cardinal number. It is not true by feel, and it is not based on spatial and temporal intuition, as Kant thought.

However, now we face another problem. The problem with this view has to do with some of the other questions about this intuition. Remember that Wright and Hale were trying to argue that we should be able to become acquainted with the concept of number through exposure to the numbers principle, and that it should not answer to some prior intuition or concept. Köhler seems to be arguing that we acquire the concept of number through this rational intuition, and this is what would justify the numbers principle as analytic.[44] Köhler also faces the bad company objection raised by Boolos. Rational intuition might justify the numbers principle, but it is not clear that it can rule out the parities principle. To do so we would have to claim that the infinity of the natural numbers is true by virtue of rational intuition, and this is why we choose the numbers principle over the parities principle. However, this argument begs the question; we need the numbers principle in order to prove the infinity of the natural numbers.

The problem runs quite deep because there are conflicting views about which is the "right" logic, in the sense of a logic that is prescriptive of all reasoning or rationality. That is, even if we accept the notion of rational intuition, there is controversy over which is the right or correct rationality and/or reasoning intuition. There are a number of formal representations of reasoning. Some differ from each other because of the subject one is reasoning about. For example, logic with temporal operators is especially designed to deal with reasoning about events happening in time. Another example is free logic, which is especially designed to help regiment our reasoning about fictional objects, or nonexistent objects of some sort. There are also formal systems of reasoning that can combine more than one of these features.

The problem has a wide scope because within each of these areas of especially tailored logics there are several formal representations, one for each combination of axioms. So, for example, there are several modal logics, several temporal logics and several free logics. The problem is that we have to choose one temporal logic over others as prescriptive of reasoning over temporal contexts. Worse, some pairs of formal systems are contradictory, and so cannot be combined. This is what we met with when considering the numbers principle and the parities principle. We cannot add both to Frege's first four laws, for we would end up with a contradiction: that the number of cardinal numbers is both infinite and finite. This problem is faced at a much more fundamental level too. We cannot agree on all sorts of extensions of Frege's first four laws.

Worse, even the first four laws are in jeopardy, for they are basic laws for creating what we call a "classical" system of logic. What we normally learn in a first course in logic are classical propositional logic and classical first-order logic. There are non-classical propositional logics and non-classical first-order logics, non-classical second-order logics and so on. What makes a logic non-classical is that it disagrees with one or more of the classical axioms.[45] We shall investigate one of these non-classical logics in Chapter 5.

8. Conclusion

We leave the logicist philosophy behind. What is central to logicism is the idea that all, or part, of mathematics is really logic. This fits perfectly with our sense of the hierarchy of knowledge, where logic appears at the top. Logic is thought of not merely as a branch of mathematics, but as setting a norm for rationality. Logic is universal, in the sense of being applicable to any area of study in a non-metaphorical way. Logic is directly applicable; we may always appeal to logic. Moreover, logic is not about anything in particular; rather, it is regimented reasoning. To argue for logicism, we need to first present a logic. The logic has to be justified in occupying this privileged place in our hierarchy of knowledge. We then have to reduce part of all of mathematics to this logic.

The three logicist groups we have discussed are Frege, Whitehead and Russell, and the neo-Fregeans. Frege's attempt to demonstrate logicism failed because a paradox was discovered in his formal logical system. This tells us that it could not possibly be logic, since logic has to at least be consistent (under classical conceptions of logic). Whitehead and Russell's attempt to demonstrate logicism was more ambitious than Frege's. They wanted to show that all of mathematics, not just arithmetic and analysis, are really logic. They failed because some of the axioms of their system are arguably not *logical* principles in the philosophically relevant sense. The neo-Fregeans try to patch up Frege's demonstration by fixing the underlying logic, but to even reduce arithmetic to logic they need the numbers principle. It turns out that there is dispute, which can be very deeply held, over whether this is a principle of logic.[46] As we can see, the study of logicism is appealing, and there is much exciting work still to be done.

9. Summary

The important points to retain from this chapter are:

- Logicism is a philosophical position that claims that some, or all, of mathematics is really logic.
- The significance of logicism is that logic is accorded a special place in our epistemology, and so logicism tries to answer the epistemological puzzles that arise against platonism.
- The logicist philosophies have been developed in considerable detail, and are quite sophisticated.
- Frege's logicism is modest: reducing arithmetic and analysis to logic. Frege made tremendous contributions to the field of logic, but his logic was faulty because inconsistent.
- Whitehead and Russell tried to reduce all of mathematics to logic. They failed because their proposed "logic" – the type theory – does not seem to be *logical* in the philosophical sense.
- Wright and Hale have recently advocated a neo-logicism, but this too is beset with problems concerning the analyticity of the numbers principle, which is meant to be a logical principle.

Chapter 4
Structuralism

1. Introduction

There are three main current exponents of structuralism: Michael Resnik, Stewart Shapiro and Geoffrey Hellman. After the introduction in this chapter, §2 is a generally motivating section on structuralism. Section 3 discusses Hellman's modal structuralism. Section 4 compares Resnik's and Shapiro's structuralist positions and §5 critiques them.

In brief, the structuralist position states that mathematics is about structures, as opposed to mathematical objects such as numbers. Roughly, a structure is a pattern. There are geometrical patterns and numerical patterns. A mathematician knows about many of these, and studies them in depth. Patterns can be complex and abstract. Visually, an example of a very simple pattern is a smooth surface of a uniform colour; one of a very complex pattern is one generated by a fractal equation.

We might think that we cannot distinguish a pattern at all when we look at, say, a landscape, but although we may not see a symmetrical pattern, we do not see complete chaos. We see a cluster of trees, a meandering brook along a valley floor, a steep hillside. As soon as we make observations, we begin to order what we see. We might notice that there are two clusters of trees, one spanning part of the brook, the other perched on the hillside. We have started counting, and we are describing the spatial relations between the things we are picking out. To discern a structure, we abstract away from much of what we see. We focus on the structural properties, and not on the content. Mathematicians are interested in structures *simpliciter*. They might apply their vast knowledge of structure to other areas of research, but this is applying mathematics, rather than doing pure mathematics.

Mathematical structures are not just collections of objects; they have as an indispensable part the relations between the objects. To commit a landscape to memory we look for patterns, and put various elements in some order, and the mathematician carries out this process at a very abstract level, and studies different types of pattern. It is important, for the mathematician, that 3 < 8. "Less than" is a relation between the two numbers 3 and 8. The natural numbers are not just a collection of objects. They form a structure: a strict ordering of the numbers by the "less than" relation. Moreover, it is an infinite structure. It has no loops, it has a beginning, and there are various mathematical ways of extending the structure into the infinite.

The structuralist position is different from traditional platonism because, for the structuralist, mathematics is not about objects *tout court*. The objects are of no importance when divested of their relations with other objects in a structure. So, the *objects of study* in mathematics are whole structures: objects together with the predicates that apply to them, relations that bear between them and functions that take us from one domain of objects to a range of other objects.[1] The objects can even be eliminated. As we shall see with Hellman's position, it is possible to construe mathematics in such a way as to do all the mathematical calculations we want without, strictly speaking, positing any mathematical objects at all. Hellman's is an eliminativist position. In a strict sense, he eliminates mathematical objects such as numbers.

We have to distinguish between what are *prima facie* objects of a theory, and concepts of a theory. Another way to put the point is to distinguish between objects of a theory and objects of study. The former are treated *as basic objects by the theory*. Objects are the base things. In physical theory, they are anything from stars, to medium-sized dry goods, to protons. Concepts such as "power", "buoyancy" and "electrical charge" are components of the theory that are applied to objects. According to the structuralist, the mathematician studies the concepts; they are the "objects of study". The *prima facie* objects of mathematics are numbers, shapes and lines. Examples of concepts are: "is a real number", "is greater than", "is a variable", "add", "is similar to" and "is an infinite set". Predicates, relations and functions are all concepts of mathematics. While these can be treated *as objects* (by virtue of grammar), they are not *base objects*. They are one level of abstraction up from objects. This is why we call them "concepts".

The platonist thinks that mathematics is about the objects: geometrical figures or numbers that are in some sort of Platonic heaven. The realist thinks that the objects of mathematics are independent of us, but are not given a location, however ethereal. For both theories, the objects of mathematics are these numbers or shapes. Moreover, the objects of study of mathematics are these numbers or shapes. The mathematician apprehends these and studies them. The tools the mathematician uses to study the objects are intuition together with a battery of concepts, such as those listed above. In contrast, the structuralist thinks of these objects as incidental. The real focus of attention is the structure. Mathematical structures are the units of interest, not the basic objects of mathematics.

Epistemologically, the structuralist is no longer concerned with how it is that we know mathematical objects but, rather, with how we know relations between objects, or how we are able to pick out a subset of objects by means of predication. It is clear that we have this ability. In fact, the structuralist thinks that it is less controversial to attribute to us this ability than it is to attribute to us the ability to know an abstract object such as a number. We shall see why when we examine the different theories of structuralism. The epistemological puzzle for the structuralist is quite different from that of the traditional realist. Since "doing mathematics" is really pattern-spotting, the slogan for the epistemology of structuralism is: "mathematical knowledge consists in the ability to spot patterns". The slogan is due to Resnik (1982). It is a good starting-point, but Shapiro and Resnik part company over the slogan. Shapiro distances himself from the pattern-spotting metaphor in order to explain, for example, infinite patterns. Shapiro points out that there is no obvious sense in which we can spot a necessarily infinite pattern, let alone tell different infinities apart. The visual metaphor of spotting is not strong enough to account for the epistemology of mathematics.

2. The motivation for structuralism: Benacerraf's puzzle

Before elaborating on structuralism, let us begin with the problem that led to it. All of the notable structuralists cited above pay homage to Paul Benacerraf's article "What Numbers Could Not Be" (1983a). Benacerraf presents a puzzle, and then hints at a solution, and it is this hint that led to the full philosophical development of structuralism.

The puzzle is this. Two different mathematical educations are contrasted. We are to imagine two children, Ernie and Johnny.[2] Each is the child of mathematicians, and each begins his education at home. Instead of being exposed to a conventional education in mathematics, they are first taught set theory: Johnny is taught Zermelo–Fraenkel set theory and Ernie is taught von Neumann set theory. The children are first taught the axioms, and then taught how to derive theorems from the axioms. This forms the base of their mathematical education. At some point both children start to receive a more conventional education, and are exposed to basic arithmetic. The children have a lot of set theory already, so they are just taught translations from words such as "number", "add" and "multiply" to their set-theoretic analogues. They are taught that conventionally one focuses on a particular infinite set, call this "numbers"; that this set has a first element, each element has a unique successor, and so on. They are taught the words to run through the series of elements of the set starting with the "first", and so on. The children can then communicate with other children who have received a more conventional education. But they know a lot more mathematics than most children and most elementary school teachers. When Ernie and Johnny meet each other and start a discussion about arithmetic, they discover that they are able to derive, and completely agree on, the Peano/Dedekind axioms for arithmetic.[3] Furthermore, the two children agree as to what counts as addition, multiplication, subtraction and division. So they agree that 2 + 9 = 11, and that 6 < 78, that 3 × 16 = 48. This also accords with what they have been taught under the more conventional part of their mathematical education. Thus, they are both very proficient at arithmetic.

Disagreement only begins at the level of ontology: at the level of the objects of mathematics. When they each try to match what are conventionally called 1, 2, 3, … to sets, one child interprets 1 to be ∅, 2 to be {∅}, 3 to be {∅, {∅}}, 4 to be {∅, {∅}, {∅, {∅}}} and so on; so each *number* is a *member* of all the ordinals that succeed it. The other interprets 1 to be ∅, but 2 to be {∅} and 3 to be {{∅}}; so each successor ordinal is encased in more set-theoretic brackets. Now 1 is a member of 2, but it is a *member of a member* of 3. Thus, for Ernie and Johnny there is a disparity concerning what the numbers 2 and 3 really are.[4] In particular, Ernie and Johnny will disagree whether 2 is a member of 6 or *a member of a member of a member of a member of* 6. This is something their conventional teacher cannot help them with, not because of lack of knowledge, but because there is nothing in our conventional wisdom concerning the numbers to decide between the two statements. Moreover, there is no absolute mathematical fact of the matter as to which is the correct interpretation of our conventional (pre-set-theoretic notion of) 6. Similarly, the traditional platonist or realist cannot help because the disagreement between Ernie and Johnny shows that the numbers of arithmetic are not just one thing, at least according to our most developed set theories. Set-theoretically, they could be one of many, quite different, things; and there seems to be no principled mathematical way to tell which is the true mathematical representation of the conventional numbers. The platonist then suffers from epistemological embarrassment.[5]

There are several reactions to this puzzle. Let us first field a platonist reaction. The more traditional platonist will stay loyal to our pre-set-theoretic notion of number. She then denies that set theory can tell us what the numbers really are. They really are 0, 1, 2, 3, … and the set-theoretic interpretation is just a representation of the numbers in a system. The problem then is to decide which set-theoretic representation is more loyal to our pre-set-theoretic views. But this is impossible, since the competing set theories agree on the all the pre-set-theoretic notions.

A more sophisticated platonist, might argue that there is one foundational set theory. This more sophisticated platonist has to battle it out with his rivals using philosophical arguments concerning each axiom, rule of inference and the semantics of the theory. This is not easily done, and the debate continues today. In particular, the sophisticated platonist has to appeal to intuitions that are not contained in our conventional learning of the numbers. Instead, the sophisticated platonist has to appeal to "fruitfulness of the theory", "applicability", its "fit" with other theories, "simplicity", "ontological parsimony" of the theory or other practical, aesthetic, metaphysical or mathematically global considerations.

To summarize, the platonist reaction to Benacerraf's puzzle is an entrenching of position, in the case of the platonist, saying that "0, 1, 2, 3, … is the essential ontology of arithmetic". Set theories are imperfect representations of essential objects. The less sophisticated platonist is left embarrassed by the puzzle, since she cannot give a philosophically principled way of preferring one interpretation, or representation, of the ordinal numbers over another. The more sophisticated platonist shifts to saying: "However, 0, 1, 2, 3, … are best represented in a particular set theory, which is a foundational discipline that gives us the essential ontology of arithmetic". The more sophisticated platonist backs one chosen set theory as being a better foundation for our theory of the ordinal numbers than our pre-set-theoretic intuitions. We then have to defend the foundational discipline without appeal to that theory, otherwise we beg the question against ourselves. Under this sort of reaction to Benacerraf, Benacerraf's puzzle is seen as a challenge to defend one foundational discipline over another.

This is not what Benacerraf himself hinted at as a solution. His suggestion is much more radical. He points out (Benacerraf 1983b) that the argument concerning which foundation is better is really a philosophical argument, not a mathematical one. Instead, he proposes that since there is no mathematical fact of the matter as to what the number 6 really is, mathematics should instead be viewed as a study of structure, as opposed to thinking of mathematics as the study of certain sorts of objects, such as ordinal numbers.

The torch is taken up by Resnik, Shapiro and Hellman. The general reaction of the structuralist to Benacerraf's puzzle is to follow Benacerraf's lead: not to hone in on one foundational theory as giving the essential objects. Instead, mathematicians study the relations between these objects as opposed to the objects themselves. They study the objects only in the sense of their occupying positions in a structure. The relations between objects are what give form to the collection of objects (Resnik 1982, 1997). Without form the objects are of no individual interest. Mathematicians study structure.

3. The philosophy of structuralism: Hellman

In *Mathematics Without Numbers* (1989), Hellman develops a modal structuralism, "modal" standing for the modes, in this case possibility and necessity.

Modal logics are developed to help us reason over possibilities.⁶ The idea behind modal structuralism is that the structures studied in mathematics are possible structures. Something is possible just in case it is allowed, by rules or concepts, or it is not impossible.

The modal structuralist thinks of mathematical theories, individuated as structures, as possibilities. A mathematical structure might or might not exist in the actual world (have an instantiation), but they are all possible. This answers directly the puzzle posed by Benacerraf. There are competing set theories, as there are competing geometrical theories. There is no purely mathematical reason to favour one over the other. There might be psychological reasons, historical reasons or reasons to do with applications within or outside mathematics, but they are not purely mathematical reasons. Mathematics does not arbitrate; it only tells us what is possible. All structures are treated on a par by the structuralist, and mathematicians *qua* mathematicians do not really care which ones people favour; which structures the actual world has is up to it. So the application of mathematics is just the recognition that a possible mathematical structure is actualized. So, the structure, in an applied case, is both possible and actual. When we apply mathematics, we apply our mathematical insights to tell us more about the "real" world. Because there is no principled mathematical way of favouring one structure over another, Hellman's approach also includes the idea that the basic objects of the mathematical structures do not literally exist, hence the title of his book *Mathematics Without Numbers*. In contrast to the position of the platonist, or traditional realist, the numbers do not exist independently of the structure in which they find themselves. Moreover, the structures do not "exist" either; the structures are just possibilities. Thus Hellman very nicely avoids all the ontological problems associated with platonism or traditional realism. He does not have to explain a "realm of abstract objects".

Instead of saying, for example, that the Peano/Dedekind axioms are true, Hellman would have us say, in our more careful moments: "It is possible that there exists a set of objects (a model) satisfying the Peano/Dedekind axioms of arithmetic". Moreover, the primitive notions used in the axioms, such as 0 and the notion of "successor", do not owe any allegiance to our pre-theoretic notions of 0 and the idea of successor. Instead, 0 and "successor" are completely defined by the axioms. A structure (composition of concepts with a domain of objects) is possible just in case the concepts work well together; that is, just in case they do not lead to contradictions. The axioms stipulate some constraints on concepts and objects. Once we have stipulated axioms, we might then try to work out if any particular progression of numbers conforms to the dictates of the axioms. If one does, then we say that the axioms are satisfied by that progression of numbers. Which particular progressions of numbers satisfy a set of axioms will be a coincidence, which is not strictly mathematically interesting. The structuralist reverses our usual way of thinking about mathematical theories. The structuralist distinguishes how we happen to come up with the theory, and what justifies the theory. For example, the way in which Peano and Dedekind came up with the Peano/Dedekind axioms was that they were trying formally to represent our intuitions about arithmetic from before we gave them formal representation. This does not justify the axioms, for once we have a good axiomatic theory to represent our former loose ideas we can leave the ideas behind. We then strictly work within the nice, new, clean theory. Our pre-formalized intuitions are, strictly speaking, a coincidence with respect to mathematics. For Hellman, the structure determined by the Peano/Dedekind axioms is justified because it is possible. Similarly, which mathematical structures best fit our physical theory (with respect to pure mathematics) will be a coincidence (Hellman 1989: 118). Nevertheless, the physical theory plays a role in stimulating and prompting mathematical developments and investigations.

But what about the structures themselves, or collections of structures? Are possible structures real? This is a metaphysical question. Distinguish between what is both possible and actual, and what is merely possible (so never comes about). Mere possibilities do not exist, according to Hellman; they are just possibilities. Hellman is an eliminativist: possible structures are not part of the ontology of structuralism. So Hellman's ontology is quite sparse.

Unfortunately for Hellman, there are different metaphysical positions concerning possibilities: whether they exist, independently of us; what their relation is to us; and so on. We shall not delve into the ideas here. Instead, they will be touched on briefly in the section on Meinongian philosophy of mathematics in Chapter 6. Here let us just register the fact that there are opposing views, and that Hellman chooses a defensible view from among them. But we might try to generate a paradox by asking the further question about gathering all the possible structures of mathematics together. Do these form a possible structure? Is this a new possible structure? Does it include itself? Hellman points out that since the structures are possibilities, "It simply makes no sense to speak of a collection … of all structures or all the items in structures *that there might be*" (2005: 556). We do not collect possibilities together, says Hellman. Metaphysically (conceptually), this makes sense,[7] especially when talking of more mundane possibilities such as possible futures. But when discussing mathematically possible structures we have to be careful and add some qualifications. It is common for mathematicians to compare structures to each other, and group structures together. For example, a mathematician might say that all of these structures share some property. The trick that the structuralist uses here is to interpret this as saying: treat structures as objects (of study). This makes the group of structures into one (meta-)structure. More specifically, when a mathematician discusses the relationship between two structures S and T, she is treating S and T as objects in a greater structure Σ. There are two objects in Σ, namely S and T, and there is a relationship between S and T. This is just a play of moving one level of abstraction up, and this sort of movement is commonplace in modern mathematics. So we are not gathering structures together, we are gathering objects (since this is how they are treated). These objects happen to be structures in their own right, but they are not, for now, being treated as such. This interpretational trick of the structuralist reflects mathematical practice well, so it is a virtue of the theory. However, we still have a paradox. This big meta-structure of all possible structures could still have itself as member, in which case it did not include all the structures, since it (necessarily) forgot itself. So Hellman has to have a block on moving to ultimate abstractions by grouping together all the structures. What prevents us from making "ultimate" abstractions? Hellman's answer is that we are blocked by the rules of set theory. Zermelo–Fraenkel set theory determines the possible structures, and a structure created by ultimate abstraction cannot arise in Zermelo–Fraenkel set theory.

However, as philosophers we want more than this; we want to justify the set theory. There are two justifications. One is that we do not want paradox and, therefore, we should not allow "ultimate" abstractions. Zermelo–Fraenkel set theory prevents these. This is philosophically not a good answer because it is *ad hoc*. The second answer is that collecting all the structures together offends against common sense. Again, philosophically, this is not good; common sense is not a reliable guide to mathematical truth. Unfortunately, although Hellman is aware of this problem he refuses to say more. This is a controversial place for Hellman to end his discussion, for we might retort that the very reason we need a philosophy of mathematics is that a lot of mathematics leaves our common sense behind. Thus, while the ontological questions are well answered by Hellman, questions about the avoidance of paradox are left hanging.

Further, it seems as though the usual epistemological questions are well treated by Hellman. We know about the structures because we study what is mathematically possible. That a particular set of axioms best fits some pre-theoretic conceptions is a coincidence, with respect to pure mathematics. Our development, or favouring, of the Peano/Dedekind axioms for representing our pre-formally expressed intuitions about arithmetic can be given a historical, or psychological, explanation. This will be discussed further in §4. The structuralist sees questions about the fit between pre-formally represented intuitions about mathematics and particular sets of axioms as lying outside pure mathematics. Our best guess as to what the limits of mathematical possibility are comes from set theory, since this is a well established and comprehensive theory.[8] The set theory championed by Hellman is the well accepted Zermelo–Fraenkel set theory. Any structure that is possible within set theory is a possible structure worthy of study. But now we can ask a more probing question: how do we know that it is Zermelo–Fraenkel set theory that sets the limits on what counts as a mathematical possibility? To answer this question "It seems we must fall back on indirect evidence pertaining to our successful practice internally and in applications, and, perhaps, the intuitive pictures and ideas we have of various structures as supporting the coherence of our concepts of them" (2005: 556). This is not a very good answer. In particular, it is awkward when we consider the motivation for structuralism, for one of the things pointed out by Benacerraf is that given a set theory there are different interpretations of structures within set theory, such as that of the "natural" numbers. The problem is compounded when we consider that there are rival set theories, with different axioms, each setting different bounds on what counts as a mathematical possibility. We wanted to avoid talking about spooky mathematical objects independent of us, but now we need them, or something like them, to justify our choice of what counts as a possible mathematical structure.

Things get worse for the modal structuralist when he tries to account for infinite structures. Under Hellman's views (1989: 80), it is possible that there should be no infinite structures. It is also possible that there should be infinite structures (*ibid.*), since it is logically possible that there should be infinite structures. This means that much of the study of mathematicians, which concerns infinite structures, is only warranted by an axiom of infinity. Such an axiom belongs to set theory, but it is independent of the other axioms, that is, we could consistently run a set theory with the axiom of infinity, or we could run a consistent set theory with no such axiom. So there is a sense in which we choose whether to include an axiom of infinity in our set theory.

There is something unsatisfying in the modal structuralist story about infinity. The modal structuralist says that it is possible that there should be infinite structures, but will say little more than this, for mathematical possibility is a "primitive" notion. This begs the question, since there is a choice, made by Hellman, of the set theory governing possibility and necessity. It remains that this is one choice among others, and that under a different choice other proposed structures, other than those recognized by Hellman's championed theory, would be recognized by the theory. In particular, Hellman could have chosen a set theory with no axiom of infinity. In this case, an infinite set would be a mere possibility (allowed, but not guaranteed, by the theory). In contrast, in Zermelo–Fraenkel set theory there is an axiom of infinity, and therefore necessarily there exists an infinite set. In another set theory that contains an axiom stipulating that all sets are finite, it will be impossible for there to exist an infinite set. At this point, Hellman appeals to the notion of a "very natural" theory. The theory he chooses is accepted by mathematicians, it has stood the test of time, and sits comfortably with the applications of mathematics in physics. Unfortunately, it falls short of answering the deeper metaphysical questions about the reality of the possible structures. Hellman recognizes this:

> As has been recognised, however, the modal-existence claims [resting on a particular set theory and philosophy of possible worlds] raise questions of their own. We see no way of explaining them away as linguistic conventions, or of otherwise reducing them to a level of observation, computation or formal manipulation. At best, the modal approach involves a trade-off *vis-à-vis* standard platonism, and is far from a final resolution of deep philosophical issues in this corner of the foundations of mathematics. (1989: 143–4)

We shall further develop the notions of choice of theory, and which structures exist, in the next section.

4. The philosophy of structuralism: Resnik and Shapiro

Shapiro and Resnik hold different positions, but share rejection of Hellman's modal structuralism. They do not need the modal notions because they think of structures as *sui generis*, that is, the structures are not further explained in terms of more primitive modal notions. Structures are not eliminated in favour of possibilities or anything else. To distinguish Resnik's and Shapiro's positions, let us return to some generalities about structuralism.

First, let us see what the structuralist says about basic objects such as the number 2. As structuralists, we can be quite precise and think of these as first-order objects.[9] There are two ways of thinking about first-order objects in structuralism, corresponding to a dispute among structuralists. One way is to think of objects in structures as *ante rem* (before reality). This is Shapiro's view. The other is to think of them as *in re* (in reality). This is a view shared by Hellman and Resnik. An *ante rem* structuralist thinks of the structures as existing quite independently of whether there exist objects that happen to exhibit the structure. An *in re* structuralist grounds the structures in applications in the real world: usually in our theory of physics. The *in re* structuralist believes that mathematical structures only exist in so far as there exist objects that have the particular structure. Which objects exist is not determined by mathematics, but by the world independent of mathematics. This is the main bone of contention between Shapiro, on the one side, and Resnik and Hellman, on the other side.

To elaborate, the debate between *ante rem* structuralists and *in re* structuralists is one level of abstraction up from an old dispute about properties, such as "is blue". A philosopher who thinks of properties as *ante rem* will maintain that colours exist quite independently of whether there are any physical objects that sport the colour. In contrast, a philosopher who thinks of properties, such as colours, as *in re* thinks that the colour only exists if there are objects in the world that have it. If there should be no physical objects with a certain colour, then that colour will not exist. For the supporter of the *ante rem* view of properties, this is counter-intuitive because some colours will come into, and drop out of, existence along with the objects that have those colours. Note that colours will also be located (in the objects which have them). So we could legitimately ask the philosopher who champions *in re* properties: "Where is the colour aquamarine today?"

This might not be as bad as it at first seems to be, since "white" light will diffract into all colours when shone through a prism or through a drop of water. Thus, all the colours are present provided there is "white" light around. However, insisting on this is not entirely helpful. For one thing, it is somewhat anachronistic. The debate between an *ante rem versus* an *in re* conception of properties took place well before Newton's famous discoveries about the diffraction of light. But even if we ignore the anachronism, we can think of the "energy-death" of the universe, in accordance with the law of entropy. At some point, as the universe approaches zero energy, there will no longer be any white light. There will only be a reddish light, and then the universe will go dark. Thus there will be some colours that had been instantiated, which will no longer be instantiated. In the *in re* properties view, colours will literally cease to exist: not just instances of the colours, but the very colours themselves. To many people this will sound implausible.

Generalizing the discussion from properties to structures, *in re* structuralists, such as Hellman or Resnik, think of structures as abstracted from objects or collections of real, or actual, objects. The objects come first. The structure is lifted from these (by abstraction). If no real objects have a certain structure, then there is no such structure. This makes a nice point about the "application of mathematics". The platonist, or traditional realist, has to account for our ability to apply mathematics to the world. The question the traditional realist has to address is: given that the objects of mathematics exist quite independently of our ability to know about them, how is it possible to bring the mathematics down to earth, to count, say, physical objects? Is the applicability of mathematics a miracle, or just good fortune? The *in re* structuralist has no such difficulty, since he simply lifts the structure from existing objects through a process of abstraction. This process of abstraction is carried out by mathematicians, who have the ability to see structures or spot patterns. This pattern-spotting is the process of abstraction. The mathematician ignores what it is she is abstracting from, and just discerns the relations between the things. The *in re* structuralist position is attractive, but it faces some problems.

The mathematical case for *in re* structuralism is also somewhat implausible under a certain assumption. The assumption is that mathematical objects exist as constructs of our minds. Thus, for example, until we had "invented" or "spotted" non-Euclidean geometric structures, the geometry did not exist. Again, the implausibility of the position lies in the thought that mathematical structures are temporal: they exist at some times and not at others. Their existence depends on our minds. If we are not thinking about a certain structure, or if we have not written down a definition of a certain sort of structure, then it does not exist. This is implausible in the light of a strong intuition that mathematics is timeless, or transcends time.[10]

However, there is a way out. The *in re* structuralist does not have to make the above assumption. The *in re* structuralist can think of mathematical objects as existing all the time, quite independently of our abilities to know that they exist. The structures that supervene[11] on the objects similarly exist independently of our knowledge of them. Unfortunately, this type of *in re* structuralism is not so philosophically robust either, for the existence of mathematical structures seems to depend on the existence of real, or actual, objects that participate in certain structures. If those objects cease to exist, then the structures do too. If new sets of objects come into existence that happen to have certain structures, then the mathematical structures come into existence. The *in re* structuralist now needs to tell us what sorts of objects can participate in a structure. If the *in re* structuralist believes that the only real or actual objects are physical objects, then only the mathematical structures exhibited by physical objects are *bona fide* mathematical structures. There are problems with this. One is that there is a lot of mathematics, about which we have no idea whether it is applicable. So the *in re* structuralist story runs against mathematical experience, which develops the structures first and worries about applications later. Another problem is that in accounting for infinite mathematical structures, it is not clear at all that there are an infinite number of physical objects around for mathematicians to abstract from, to legitimize their development of theories of infinite numbers. Another problem arises when we think of the "death of the universe". It is not clear that when the universe dies there will be any physical objects around at all, since there will be no energy to bind atoms and molecules. There will be no physical structure at all, or at best there will be one structure. Again, the problem is that mathematical structures end up being temporally constrained; they come into existence and go out of existence with the coming into existence and going out of existence of the physical universe.

In re structuralists do not have to confine themselves to physical objects having a structure. We can count ideas, or abstract objects, or follow Hellman and discuss possible structures, but then we are back where we started. We wanted to avoid talking about spooky possible mathematical objects independent of us, but now we again need them, or something like them, to justify our discussions of infinite structures. Viewed this way, the *in re* position is self-defeating. The structuralist was supposed to help us navigate around the philosophical problems associated with realism in mathematics, and instead has landed us squarely in the problems.

Shapiro's *ante rem* structuralist tells a different story. The *ante rem* structuralist tells us that structures exist independently of any objects' "being there" to instantiate the structures. The idea is that mathematicians know about, and study, some structures. The mathematician studies not individual mathematical objects, but rather sets of objects with relations that bear between the objects. There is no interest for the mathematician in studying the number 8 in isolation from the other numbers.[12] The mathematical object 8 is nothing more than, and nothing less than, a place in a structure; the structure is what is mathematically important; 8 by itself is incidental. The "place in a structure" might fail to be occupied by anything at all. Whether a position in a structure is occupied is of no interest to the mathematician.

There is a puzzle. The number 8 figures in several structures: the structure of the natural numbers; the structure of the integers; the structure of the rational numbers; and so on. We might ask the structuralist how it is that the "same" object (place in a structure) should belong to different structures. The answer has two stages: the diagnosis and the remedy. The diagnosis is that there is a historical account of why the "same" number occupies different structures. We started by thinking about the natural numbers, from these we developed the integers, from these we developed the rational numbers, the real numbers and so on. Historically, one structure of numbers arose out of another. Each is a proper subset of the other. There is a sloppiness in language that we can easily tidy up, to properly understand locutions such as "8 is a member of both the natural number structure and the integer structure". The sloppy thinking can be overcome by taking the remedy. In the remedy we understand that there is a meta-perspective from which "'8' in the natural numbers" is equivalent to "'8' in the integers". This is the perspective that says that natural numbers are a proper subset of the integers. From another meta-perspective "'8' in the natural numbers" is equivalent to "'8' in the real numbers". In the interest of being precise we then discuss "equivalence of numbers" rather than "identity of numbers", since the structures are very different. Diametrically, there are meta-perspectives in which the "8" of the real numbers is quite different from the "8" of the integers.

Let us elaborate on the historical diagnosis. "Conceptual history" is deliberately ambiguous between the individual human "learning about the number 8" and how the history of mathematics gives us different perspectives on the number 8, as we progress in our collective, or "best human", mathematical conceptions. The *ante rem* structuralist gives both an individual learning diagnosis and a historical collective diagnosis.

The individual learning aspect of the conceptual history starts with observing children learning mathematical concepts. Children, not born of mathematical parents *à la* Benacerraf, begin learning their mathematics by learning to recite positive whole numbers by rote, then to count small finite quantities of objects. The structuralist identifies this stage of learning with learning about numbers as objects. Counting is made possible by understanding a relation between numbers (as objects) and groups of, say, physical objects. As the child learns that counting can continue indefinitely, and learns to add and subtract numbers from each other, and then multiply and divide numbers; the child's understanding becomes more sophisticated. The numbers take on a meaning of their own that no longer depends on the presence of physical objects. One number can be subtracted from another number, without any evocation of physical objects. At this stage in learning, the child has started to think structurally. Numbers are places in a structure: the structure of the natural numbers.

The child will then move on to the next stage, and will learn about negative numbers and other structures: rational numbers, real numbers and so on. The structuralist then says that the child learns that there are other structures, but still each is compared to the other in an informal way. Each structure is presented individually. So, "'8' in our integer structure is the same '8' as the '8' in the natural number structure" is only informally correct.

The more sophisticated stage, usually reached only in adulthood, if at all, concerns comparing arbitrary structures, as in model theory. The model theorist leaves the natural numbers behind altogether. The model theorist is interested in there being some set that satisfies, or models, a structure, because that guarantees consistency, but does not care which sets model the structure. The real interest for the model theorist is to compare structures to each other, or structure types (groupings of features of structures) to each other. For example, a mathematician might be interested in effective structures. In this case he says something like the following: "The objects of the study of effective mathematics are the effective mathematical structures. A structure is effective if its universe is computable and its operations and predicates are uniformly computable" (Dimitrov 2002: 1).[13] At this stage, "8" is only the same "8" in "different" structures, if it satisfies a number of mathematical properties. For example, in several structures "8" is the immediate predecessor of "9". This will be true of the whole number structure, or some finite subsets of it, and in the integer structure. It will be false in the structure of the rational numbers, where "8" has neither an immediate successor nor immediate predecessor because of the density of the rational numbers.

Now we turn our attention to another interpretation of the history. The *ante rem* structuralist can tell a story about the collective human study of mathematics. Human beings began studying particular problems in mathematics and developed techniques for solving the problems, and then the problems were applied elsewhere. Babylonian and ancient Egyptian mathematics consisted in formulas for calculating how many bricks were needed to construct a ramp of a certain size and how many loaves of bread were needed to feed a certain number of slaves. The Babylonians had tables for "squares, cubes, square roots, cube roots and even roots for the equation $x^2 (x + 1) = a$" (Anglin & Lambeck 1995: 22). Neither the Babylonians nor the ancient Egyptians had proofs for their tables or formulas. It was the ancient Greeks who introduced the notion of proof, and thereby not only a general way of generating a table,[14] but also a justification for the figures in a given table.

Plausibly, the structuralist thinks of Babylonian and Egyptian mathematics as akin to the young child learning a few particular finite numbers. Being able to give proofs requires a higher degree of sophistication than being able to apply a formula or look up answers to a problem in a given table.[15] Once we have a sense of proof, we have moved up a level of sophistication and abstraction. We then have the conceptual tools to dispense with tables and to deal with arbitrary situations. Our "mathematical" thinking changes from being able to solve applied problems concerning loaves of bread, to reasoning in general.

It was not until the twentieth century that we saw the emergence of model theory as a mathematical discipline. This is a mark of conceptual maturity for Shapiro's version of structuralism. Or, more precisely, Shapiro proposes to give a philosophical underpinning to this stage of our mathematical understanding.

We need first to get a feel for model theory. Model theory is sometimes called "meta-logic". Some model theorists refer to themselves as logicians. The model theorist is interested in characterizing structures according to their mathematical properties. The model theorist will also compare structures to each other. Classical results in model theory include showing that a structure, such as propositional logic, is sound or complete, has the downward Löwenheim–Skolem property, is compact, is decidable and so on.[16] Other structures might lack these properties. Comparing the properties of one structure to another is the stuff of model theory. The model theorist does the mathematics of the structuralist philosophy of mathematics. So, the model theoretical perspective gives the remedy to the problem of why it is that the same number "8" appears in different structures. The model-theoretic answer is that it is only recognized as the same number "8" in certain meta-structures.

If one follows Shapiro, and champions an *ante rem* structuralism, then any objects can fill the places in a mathematical structure. This also explains, rather nicely, the applicability of mathematics to anything at all. This general applicability is sometimes referred to as "topic neutrality" or as the "universality" of mathematics.[17] Shapiro uses the term "free-standing". Mathematical structures (as opposed to other sorts of structure) are free-standing. A free-standing structure implies universal applicability and no ontological commitment to first-order objects.

To summarize, the Shapiro-fashioned structuralist rejects Hellman's modal primitives and Resnik's *in re* view of mathematical objects in structures. Instead, Shapiro champions model theory as the branch of mathematics that best describes mathematics. The essence of mathematical activity is seen, by the structuralist, as an exercise in comparing mathematical structures to each other. This is how the structuralist characterizes current sophisticated mathematics. To labour the point, the Shapiro-inspired structuralist thinks of model theory as the canonical mathematical theory of the philosophical position of structuralism. This easily solves the problem of "the same number '8'" appearing in different structures. The judgement as to whether the symbol "8" refers to the "same place" in different structures depends on a meta-structure. That is, the judgement necessarily takes place from a meta-perspective. The structuralist is explicit about this. "The '8' of the natural numbers is the same '8' as that in the integers" is true under certain properties of structure. The "certain properties" are ones that can be recognized explicitly from a meta-structure. Both "8"s are the immediate predecessor of "9"; both "8"s are divisible (without remainder) by 1, 2 and 4. By choosing our meta-structure, we decide which properties, relations and functions to consider. If we are interested in the concept of "immediate predecessor" then in respect of that property, the "8" of the integers is different from the "8" of the real numbers, which has no immediate predecessor. This draws out the difference between equivalence and identity quite well. The "8"s in different structures are not identical to each other; they are equivalent. That is, they share some characteristics.

5. Critique

With every philosophy of mathematics, a very natural question to ask is: what are the ontological commitments of the theory? The structuralist distances herself from the traditional realist position precisely with respect to the matter of ontological commitment. Recall that the traditional realist is committed to the existence of the basic, first-order objects of mathematics. So all the numbers exist, all the shapes exist and so on, unless the realist reduces mathematics to a founding discipline, in which case the realist is committed to the existence of the objects recognized by that discipline, and no other. For example, the realist who shows that all of mathematics is reducible to set theory will have all the sets of the set-theoretic universe as the ontology of mathematics.[18] In the case of set theory this is quite neat, since the set-theoretic universe is "constructed out of" the empty set; we get the ontology of mathematics *ex nihilo*. Explaining this coherently, and philosophically, is what is difficult with the set-theoretic realist position, especially with respect to the axioms of set theory, which are independent and controvertible, such as the axiom of choice or the axiom of infinity. Hellman's eliminativist position eliminates the objects of the traditional realist. The objects are eliminated in favour of the modal notions concerning possible objects and possible structures. This elimination has its own philosophical problems, for we want to know the ontological status of the possibilities. Are possible structures real? Do they exist independently of us? How do we know about them? We saw Hellman's reply in the quotation; he does not answer these "deep questions". So we seem to have just shifted the aim of the problems faced by the traditional realist from first-order objects to possible structures.

The *ante rem* structuralist is not committed to an ontology in this sense. The objects of mathematical study are neither the elements of sets nor what are labelled "objects" in a given discipline of mathematics, nor are they possibilities. Instead, the *ante rem* structuralist claims that the objects of study of mathematics are structures. What the mathematician labels an "object" in her discipline, is called "a place in a structure" by the structuralist. The philosophical significance of this translation is that there is no ontological commitment to what the mathematician labels "objects". What the traditional realist thinks of as an "object" might or might not exist, according to the structuralist; and, frankly, this is just not an important question for the philosophy of mathematics. This is very much in keeping with Benacerraf's hinted position.

The *ante rem* structuralist stance towards mathematical ontology has two great advantages. One advantage the structuralist claims over the traditional realist is that he does not have to give an epistemological account of spooky, ethereal, timeless objects. Rather, the structuralist simply says that mathematical structures can be applied to any objects one wants. Thus, we can use mathematics to count tables and chairs. We can also count ideas or mathematical object or concepts. Whether chairs and ideas exist is a matter of no concern to either the mathematician or the structuralist. The second great advantage is related to the first. It is that the applicability of mathematics is also not mysterious or, at least, it is not something that the structuralist has to explain. This is an empty question, because all we should do is observe that we do happen to apply mathematics. This is a descriptive claim; we cannot transform it into a normative or prescriptive claim. Put another way, it is a matter for metaphysics to explain how it is that mathematics is applicable to the physical world, or the imaginary world. In contrast, the philosopher of mathematics, under the structuralist conception, should occupy himself with mathematical activity, not with how mathematics can be used.

To repeat, the "object" in a mathematical theory is a place in a structure. The structure is what is salient; it is this that is going to have importance for the philosophy of mathematics. In keeping with this thought, we can ask the structuralist whether structures exist independently of us. The answer is that they do. We can count structures that *have* certain features; we can relate two or more structures from the perspective of a meta-language or a meta-structure; we can arrange structures into equivalence classes: thus, we can count structures. When we do this we treat them as objects. Are structures objects in the thin sense of "the mathematician does not care, and need not care"? Or do structures exist in the "thick sense", that is, in the metaphysical sense of being objects that exist, and to which we ought to be committed?

We have moved the question of ontological commitment up one level of abstraction. We are no longer asking whether the (what the mathematician calls) objects of mathematics, are objects, in the "thick" or "robust" sense of counting as objects in their own right. This question lies outside mathematics for the structuralist. Rather, we are asking about the ontology of structures. Do structures exist, in the "thick" or "robust" sense? Philosophers can go two ways on this: the anti-realist way and the realist way. Shapiro opts for the realist stance towards structures of mathematics.

The realist structuralist claims that structures exist;[19] they are not inventions, or creations. Mathematical structures exist independently of us; we discover them. We can now ask the standard questions: how many of them are there? Could we be quite mistaken about them? Where, or how, do they exist?

Let us begin with the question of how many of them there are. Since real structures neither drop out of existence, nor do we seem to have exhausted the possibilities of mathematical structure, there does not seem to be a finite mathematical ceiling to the number of structures. We seem to be discovering "new to us" structures. We have to be careful in answering this, for we want to avoid the paradox involved in asking about the structure of all the structures. Shapiro cleverly does not allow unrestricted quantification over structures.[20] That is, we cannot talk about "all the structures". Similarly, we may not simply ask how many structures there are. We have to be more precise. We have to ask questions such as: if we use this meta-structure from which to assess the number of structures, then how many structures are there? The answer is always in conditional form. Cardinality depends on which meta-perspective we adopt, because that will tell us how to individuate, or pick out, structures.

We can now start probing quite deeply, philosophically. So far we have been rather glib about the notion of counting structures, or identifying, or individuating, a structure. Recall that to identify a structure we need to know when what we thought were two separate structures turn out to be one.[21] How we identify, or count, structures will, as with the "how many" case, depend on our perspective. A perspective is what allows us to discern some features and ignore others. Is there a "God's-eye" structure, from which we can "see" all structures? Given the discussion above about the number of structures, the answer should be fairly obvious: there is no unique major structure. Instead, we have to work piecemeal. That is, if we want to know whether what we think are two structures is really one structure, then we have to adopt a perspective from which to make this judgement. From a different perspective we could come up with a different answer. The answer is conditional, not absolute. This sort of answer should be familiar from our discussion above of the number 8.

The second question above was whether we could be mistaken about them. In Chapter 3 we saw one sense in which we could. We might think we are discovering, or developing, a structure, and then realize that it is deeply flawed, in the sense of being inconsistent. This is one sense in which our methods of investigation are imperfect. The structuralist is not worried about this because this is in keeping with mathematical practice. We learn that structures do not exist, when we learn that they are inconsistent. We do not know this in advance of developing, or trying to discover, a structure, and we could be mistaken about the structures we study today.

A well-developed structuralism is quite a robust philosophical position, since it answers Benacerraf's puzzle, and some other pressing philosophical questions. However a deep problem remains concerning the choice of underlying mathematical theory to pick out the structures and compare them to each other. Zermelo–Fraenkel set theory is not a good choice, especially when coupled with modal notions, for both modal notions and Zermelo–Fraenkel set theory are beset with philosophical difficulties, one with metaphysical questions about possible worlds, the other with questions about foundations of mathematics. Shapiro's choice of model theory is more cunning. Model theory is not a foundation, in the way that set theory is. In particular, model theory is not an axiomatized theory. There is no "ontology" of model theory. Model theory is better thought of as a perspective on mathematics. So in one fell swoop we avoid the problems we encountered above.

The model-theory perspective is a sort of organizational perspective, which organizes mathematics into structures. We examine one structure from the perspective of another (meta-)structure; we compare two structures from the perspective of a third meta-structure: it is structures all the way up. This gives a good global perspective on current mathematical thinking.

To criticize this position, we have to be equally cunning. We can complain about the temporality of the theory. That is, it is a good theory for current mathematics, but not for past mathematics (see the historical diagnosis). It might not be a good theory for future mathematics either. Mathematicians might start to favour a different perspective. The cracks are already showing, for mathematicians will sometimes discuss issues that they consider to be mathematical but that are not recognized in model theory.[22] More precisely, model theory is written in a formal language. The formal language has a certain, very high, expressive power, that is, we can represent loyally, in the language of model theory, most mathematical concepts. However, we cannot represent all of them. Details are quite abstruse, so the reader will be spared them here.[23] The criticism against structuralism is quite general. It is that, structuralism has to be underpinned by some mathematical theory, which might, or might not, be considered to be foundational; and that mathematical theory will leave some mathematics out. When it does this, then the structuralist theory is not describing all of mathematics, so it is not a philosophy of all of mathematics but, rather, a philosophical theory of some very large part of mathematics.

6. Summary

The important ideas to retain from this chapter are:

- Structuralism's major claim is that mathematics is about structures, not first-order objects.
- First-order objects are places in a structure.
- One can be a realist about structures or an anti-realist.
- The deepest problems with structuralism concern not paradox, but underlying mathematics or logic.

Chapter 5
Constructivism

1. Introduction

To further understand realism, it is useful to contrast it to its opposite position: anti-realism. Most of the philosophical positions in this book were developed as reactions to problems concerning platonism. The reactions can be moderate, as when we modify platonism slightly, but the constructivist is an anti-realist and, as such, wholly rejects the realist thesis that mathematical objects, or mathematical truths, are independent of us. Instead, for the constructivist mathematics is fundamentally a construction of our minds. We do not discover mathematical truths or objects; we construct them. The constructivist also thinks that our minds have tried to construct quite fabulous objects, but that they are not legitimate fabrications. The process of coming up with them was misguided. To continue the metaphor, some of the "constructions" of the realist are flawed. The constructivist urges a rethinking of the foundations of construction for mathematics. In particular, the logic guiding the construction of mathematical objects has to be epistemically constrained. We shall see that the constructivist thinks quite differently from the realist; he has a different underlying logic from the realist. The logic acts as a different norm of reasoning from classical logic. Constructivists share the feature that they want to revise classical logic.

We have so far been using the terms "anti-realist" and "constructivist" interchangeably, but will now clarify the vocabulary. "Anti-realist" is a general philosophical term. One can be an anti-realist in ethics, science, language and so on. Constructivists are anti-realists specifically about mathematics. To confuse the issue further, the term used historically for "constructivist" was "intuitionist", introduced by L. E. J. Brouwer at the beginning of the twentieth century. Intuitionism is a certain specialized sort of constructivism. There are many constructivist positions. What distinguish them from each other are the different underlying logics that they use to marshal their reasoning. For reasons of historical and philosophical importance, we shall focus more on intuitionism than on the other constructivist positions. In particular, we shall look at an exposition of intuitionist logic to get a feel for the different logic, together with the sorts of argument summoned in favour of the logic. This should prepare the reader for delving into the constructivist literature.

The realist considers classical logic to set the standard for correct reasoning. Recall that classical logic can be characterized by the following features:

- the law of excluded middle holds;
- double negation elimination holds;
- *reductio ad absurdum* proofs are all fine; purely existential proofs are all fine;
- and the axiom of choice holds in full generality.

Let us briefly revise these terms. The law of excluded middle says that for any formula either it or its negation holds. Double negation elimination is the rule that says that from a doubly negated formula, the un-negated formula follows (in classical logic, p can be derived from not-not-p; not so, in constructive logics). *Reductio ad absurdum* proofs are ones which proceed from assuming the opposite of the conclusion, proving a contradiction, and then rejecting the assumption, by asserting its opposite. A purely existential proof is one that proves that some thing, or number, exists, but gives no way of finding an example of such a thing or number. The axiom of choice says that every set has a "representative member" (a sort of arbitrary member). More generally, from a collection of sets, one can always assume that there is a member of each, which we can take out and put into a "choice set of members". This will be more obviously true if the members are all quite similar to each other. It is not so obvious if they are all quite different from each other. For example, take the odd set $\{0, \aleph_1, 34/6\}$. For the classical logician, odd combinations of objects do not matter; it is enough that, for the objects to be gathered into a set at all, there has to be something "common" to all of them (which might be "happens to be a member of this odd set"). The choice member is guaranteed by what we call a "choice function". We know that such a choice member exists (stipulated by the axiom), although we might not know how to pick each representative. Many of these ideas are deeply related to each other, and we shall see this later. We shall also discuss some of these features of classical logic in detail, to see how the constructivist rejects them.

On a metaphysical level, a realist about some discourse is someone who believes that the truths of that discourse are independent of us. We have a remarkably good ability to track those truths. We cannot learn all the truths there are but we collectively (the human race) shall learn a fraction of the truths out there. We refer to a truth or falsity simply by articulating a fact in a grammatical way, by means of a sensible declarative sentence. Here "sensible" just means "without any categorical mistakes": in the sense of attributing something to an object that it is inappropriate to attribute the object. For example, a perfectly grammatical nonsensical sentence is: "The colour of the number 8 elaborates motionlessly". It makes no sense to attribute a colour to a number, let alone to suggest that a colour elaborates anything, nor does it makes any sense to say that elaboration can be motionless. Unless one is involved in an extremely strained metaphor, these are all categorical mistakes. In other words, every declarative sentence that makes sense does so in virtue of being both grammatical and not making categorical errors. For the realist, the lovely feature about sensible declarative sentences is that they are true or false, independently of our ability to know whether they are true or false. We say that a sensible declarative sentence is "truth-apt", that is, it has a truth-value. So the realist says that a grammatical declarative sentence that does not make category mistakes has met all the criteria necessary for making the sentence truth-apt. This implies that there exist sentences the truth of which we shall never know, or could never know. These are referred to as "verification-transcendent truths"; they transcend our abilities to verify them. The realist is committed to saying that the following sentences all have a truth-value, despite the fact that, for all we know, it is impossible for us to know what the truth-value is:

- There is a collection of eight stones forming a heap on the dark side of the third moon of Jupiter;
- The last dinosaur was female;
- The universe is expanding in the sense of everything becoming proportionately bigger, together with gravity and the other laws of physics conspiring to make the expansion undetectable to us;
- The world was created eight minutes ago, but we cannot detect this, since it was created complete with a history;
- This particular joke X is very funny despite the fact that no one has ever, or will ever, laugh at it;
- There exist entities that cannot causally interact with us, but nevertheless have spatial location;
- There are an infinite number of prime pairs.[1]

Some of these sentences are more plausibly truth-apt than others. These are the ones that one is inclined to be a realist about.

In contrast, an anti-realist ties the notion of truth to the notion of knowledge or, rather, possible knowledge: knowledge "in principle". The anti-realist thinks that it makes no sense to entertain the idea that a sentence might be true, quite independently of our ability to know whether it is true. The anti-realist thinks it is not rational to say that the above sentences must have a truth-value when we cannot, in principle, know what that truth-value is. For the anti-realist, truth belongs to us, it is our servant, and as such, it must be "epistemically constrained". Epistemic constraint is added to the requirements of being grammatical and categorically correct for truth-aptness of a declarative sentence. For the anti-realist, all three constraints must be met for the sentence to be "truth-apt". The epistemic constraint is strong. We have to have some idea of what it would take, or what it would be like, for the above sentences to be true, or for the above sentences to be false. For the anti-realist, it makes no sense to speculate about things we cannot, in principle, know about.[2] The anti-realist draws a distinction between what is wholly impossible to know, and what it is simply impractical to know. It might be impractical for me to know all of the factors of some large number, just because it would take me a long time to work them out. This is different from "impossible" to know, where even if I had the time and motivation I could not know.

We should be a little more careful. People, even philosophers, tend to be realists about some things and anti-realists about others. As such, a person or philosopher might allow some, but not all, of the example sentences. Which ones he allows will depend on what he thinks he can in principle know. The point is that for the anti-realist (about subject x) there is no class of sentences, in the subject x, that make sense and yet we cannot know, even in principle, what their truth-values are.

The reason the anti-realist ties truth to knowledge is that he thinks that we do not innocently "track" reality (as the realist thinks); rather, says the anti-realist, we construct reality before us.[3] What is true or false depends partly on us and how we approach the world, what we choose to pick out, what we choose to focus on, and this brings in Kantian philosophy. Kant was an idealist, which is a type of anti-realism. For him "the world" is partly of our making. More specifically, he believed that there is a noumenal world and a phenomenal world. The noumenal world is what lies beyond us; it is the world, as it is, independent of us. We are part of, and are situated in, the noumenal world. However, when we interact with it, or even just perceive it passively, when we have experiences, we do so not directly but indirectly through our concepts. The result of our concept-influenced way of interacting with the noumenal world is the phenomenal world, which is as close as we can get to the noumenal world. The phenomenal world is intelligible to us. This is our world, the world we bump up against, and the one we reason about. Included in this, we might also say, strictly going beyond Kant, that truths concern the phenomenal world, not the noumenal world. We can say very little about the noumenal world, except that it exists and contributes to the phenomenal world. We cannot, in principle, know what the noumenal world is like. A precondition of our experiencing at all is that we bring our way of perceiving, or our way of experiencing, with us to the experience. Strictly speaking, then, the objects of perception, or of experience, are not independent of us: they are partly produced by us.

Expressing this more tightly in terms of truth, we say that a mathematical realist claims that there are mathematical truths that are true independently of our knowing them to be true. Recall our criteria for truth-aptness for the realist: grammar and categoricity. In mathematics, provided a string of symbols is well-formed (the stipulations as to what counts as well-formed in logic or mathematics incorporates both grammar and categoricity, thanks to Russell), it is truth-apt. Particularly interesting are sentences of the form: "There exists a number with the properties, F, G and H". This sentence is truth-apt, because it is well-formed. However, we might have no idea of how to find a number with the properties, or a proof telling us that it is impossible to find a number with those properties. We have faith that our mathematical language has all the truth-apt sentences, and that our methods of proof are quite good, and tell us, of the class of sentences, which mathematical sentences are true and which are false. In particular, the sentence "There is an infinite number of prime pairs" is either true or false. We have no way of knowing whether the truth-value of the sentence is "true" or "false"; nevertheless, it has a truth-value. In contrast, the constructivist is more cautious about which sentences in mathematics are truth-apt, and which are not. The constructivist wants to epistemically constrain truth. Truth is conferred by means of a *constructively acceptable proof*. The talk of "epistemically constrained truth" comes after Kant. We might think of the introduction of this talk as a "semantic interpretation of Kant's position". The semantic interpretation says that, for the Kantian, we reason over the phenomenal world, reasoning is propositional (so concerns truths and falsehoods), and since this involves the phenomenal world, truth has to be epistemically constrained.

Kant himself did not suggest intuitionism or constructivism in mathematics. Rather, more carefully, we should say that Kant's philosophy is one of the motivations for a constructive approach to mathematics. The constructivist uses Kant's philosophy to suggest that, in mathematics in particular, we should conscientiously display the fact that we are discussing the phenomenal world of mathematics. We should reject the realist view that we reason about a noumenal world of mathematics.

In the philosophy of mathematics the most famous constructivists are intuitionists. They have a particular intuitionist logic that they think should guide reasoning in mathematics. The logic places epistemic constraints on proofs. So unless a well-formed formula is intuitionistically *provable*, or *probably* generates a contradiction, it is not truth-apt. There are other constructive logics, each rejecting some aspect of classical logic. Each constructive logic is a candidate for guiding our reasoning in mathematics. In §2 we shall concentrate on intuitionist logic. This will give a feel for the type of reasoning the intuitionist urges on the mathematician. The change is quite profound. Section 3 concerns some *prima facie* motivations for constructivism. Section 4 concerns deeper motivations, and §5 returns to the logic, namely the semantics of intuitionist logic.

To sound an early alert, the main complaint against the constructivist is that he advocates too radical a revision of mathematical thinking. The mathematician who uses classical logic sees the revision as a rejection of too much good work. If we are really to revise our way of thinking in mathematics, then there are many results that are not acceptable and have to be rejected. The constructivist says that we really do not know whether they are true or not. To the realist, this just seems perverse, since we have a perfectly good classical proof of the result. Too much of mathematics is rejected by the constructivist, and this is unpalatable to the realist.

2. Intuitionist logic

The idea behind the logic is to rule out the law of excluded middle: for any well-formed formula A, either A or not-A holds. In classical logic this is intimately related to the law of bivalence, which says that there are only two truth-values, true and false, and every well-formed formula A is either true or false, and not both. Once we reject the law of excluded middle, we find that the law of bivalence is quite independent. That is, we can accept it or reject it. We can reject it in different ways: by adding another truth-value; by allowing sentences no truth-value; and by allowing sentences more than one truth-value (to be both true and false, in the case of paradoxical sentences). Each of these strategies can be found in some constructivist theory or other. An intuitionist, typically, retains bivalence, but rejects the law of excluded middle. Moreover, rejecting the law of excluded middle has repercussions for a number of simple rules of inference, such as double negation elimination, *reductio ad absurdum* and *modus tollens*. The last two are not rejected outright, but they are more carefully expressed than in classical logic. Furthermore, many axioms of formal systems are either rephrased or rejected outright. These include the axiom of choice and the axiom guaranteeing the existence of infinite sets. The semantics of intuitionist logic is different from that of classical logic. Thus the understanding of the connectives is different. We shall say how in detail in §5. However, first, we should discuss each of the rules of classical logic that are dispensed with or reformulated.

We begin by recalling two basic facts about classical logic. First, sentences, or well-formed formulas, in logic are either true or false, and not both. Secondly, an inference, or argument, is valid if and only if whenever the premises are true, so is the conclusion. If we want to justify rules of inference, we do so in terms of the universal validity of the rule: that it preserves truth from one formula to another. When we make this justification, we have to draw on a pre-formal notion of truth: universal validity. Otherwise we would beg the question. Turning to our system, let us look at the simplest rule: double negation elimination. This is simply the rule that if a well-formed formula is double negated, then we can infer the un-negated version of the formula. For example, where A is a well-formed formula, we can infer A from not-not-A. In symbols: $\sim\sim A \vdash A$. The intuitionist rejects this altogether. Remember that truth is epistemically constrained. So asserting A is not just asserting that it is true, in the classical sense, but also asserting that we know A to be true. Turning to double negation, if we do not know that we do not know A, it does not follow that we know A. So the inference above is not intuitionistically valid.

In contrast, double negation introduction is allowed in intuitionist logic. Double negation introduction in classical logic is the rule that if we have A, we may infer the double negation of A. In symbols: $A \vdash \sim\sim A$. This is allowed by the intuitionist because if we rephrase it in terms of knowledge, then it sounds plausible. If we know that A, then it follows that we know that it is not the case that we do not know that A. We shall not have this as a simple rule, but as a special case of *ex falso quod libet*. "Knowing X" means that we have a proof for X, or know how to generate a proof. To make this thinking explicit, the intuitionist reformulates negation in terms of proving a contradiction, symbolized "Λ" (Greek "lambda"). If $A \vdash \Lambda$ then $\sim A$. The single negation introduction rule for the intuitionist echoes the *reductio ad absurdum* rule of classical logic. It says that if from the negation of A, you can prove a contradiction, then $\sim\sim A$. We shall switch to the symbol "$\neg$" for intuitionist negation (because the meaning of the connective is really quite different in intuitionist logic). We shall also use a line between premises and conclusion for intuitionist deductive proof, and ":" for "there is an intuitionistically acceptable proof from the formulas on the left to the formula on the right". We use uppercase Greek letters – such as Γ, Δ, Θ, but not Λ – to symbolize sets of premises, or sets of formulas in the proof; which allows us to generalize the rule when we have a proof with many premises. The essence of the rules concerns everything but the uppercase Greek letters. In symbols, the intuitionist rule for single negation introduction to generate a double negation is:

$$\frac{\Gamma, \neg A : \Lambda}{\neg\neg A}$$

This is a special case of the more elegant rule for single negation introduction, which echoes *reductio ad absurdum* proofs:

$$\frac{\Gamma, A : \Lambda}{\neg A}$$

In words: "If from A, plus a collection of other premises, Γ, we can prove a contradiction, then we may assert the negation of A". Note that the notion of negation is directly tied to that of proving a contradiction rather than to falsity. This way of expressing the rules makes explicit how to read the rules, and to understand the thinking. Drawing out the notion of contradiction, we can re-express the rule for negation introduction as follows.[4]

$$\frac{\Gamma, A : B \quad \Delta, A : \neg B}{\Gamma, \Delta : \neg A}$$

That is, if A, possibly together with a collection of other formulas Γ can prove B and, quite separately, A together with some formulas Δ can prove not-B, then we may constructively write on a new line that Γ and Δ can prove $\neg A$. This is because A is instrumental in proving a contradiction.

This can be simplified with the more elegant rule for negation elimination, which is:

$$\frac{\Gamma : A \quad \Delta : \neg A}{\Gamma, \Delta : B}$$

That is, we have *ex falso quod libet*: from a contradiction (logical falsity) anything follows. This principle is rejected by some constructivist logics, but it is not rejected by intuitionists (see the Appendix for details). As we noted before, there is no straightforward rule for double negation elimination, which is usually used in conjunction with *reductio ad absurdum* proofs. This is because if we lack a proof of A we cannot then infer that we cannot have a proof of A proving a contradiction. This brings us directly to the law of excluded middle and bivalence.

The law of excluded middle is purely syntactic: it says that for any well-formed formula A, $A \vee \sim A$. Notice that we have used a different negation symbol: "$\sim$". This is because this law holds only in classical logic, not in intuitionist logic. We have to be careful. The law of excluded middle does not mean that we can *prove* either A or not-A; only *that* A or not-A. Classical logicians use this rule in proofs. Note that the law of excluded middle is not a semantic law. It does not say that either A is true or A is false. The semantic version of the law of excluded middle is the law of bivalence: for any well-formed formula A, either A is true or A is false. In classical logic there is no third truth-value, such as "unknown" or "undecided". In classical logic, the law of bivalence can be thought of as a semantic *interpretation* of the law of excluded middle. The only difference is that one is semantic and the other syntactic. In terms of proofs, or information, in classical logic, we can slide from one to the other with impunity. This is because both propositional logic and classical first-order logic are sound and complete.[5] The semantic proofs (through truth-tables or semantic trees) perfectly match the syntactical (natural deduction) proofs. For every semantic proof there is a syntactic proof, and *vice versa*.

In contrast, in intuitionist logic bivalence holds while the law of excluded middle will not generally hold. So, the law of excluded middle is not a "law" of intuitionist logic, because we know that we cannot have a proof, or proof of contradiction from, every well-formed formula.[6] "True" is interpreted to mean "we are able to prove"; "false" is interpreted to mean "we have a proof from this to a contradiction". Proofs have to be intuitionistically acceptable, and each rule of inference of the intuitionist system of proof is justified in terms of vindicating our three criteria for truth-aptness.

The rules for conjunction are the same as those in classical logic. Conjunction introduction is:

$$\frac{\Gamma : A \quad \Delta : B}{\Gamma, \Delta : A \wedge B}$$

In words, "If from some set of well-formed formulas Γ we can prove A, and from some set of formulas Δ we can prove B, then from those sets of formulas we can prove the conjunction of A and B".

Conjunction elimination is:

$$\frac{\Gamma : A \wedge B}{\Gamma : A}$$

In words, "If from a set of well-formed formulas Γ we can prove the conjunction $A \wedge B$, then from that set we can prove one of the conjuncts". Strictly speaking, we should specify that we can just as well prove the right conjunct as the left, since conjunction is commutative. Explicitly:

$$\frac{\Gamma : A \wedge B}{\Gamma : B}$$

is also a possible rule of conjunction elimination.

We can prove the commutativity of $\wedge$, so (ignoring the last rule, for deriving B) we notice that we have a pair of rules for $\wedge$: an "introduction rule" and an "elimination rule". Most of the rules in intuitionist logic can be grouped in symmetrical pairs like this. The thinking behind the choice of rules is that the introduction rules allow us to add a symbol to a formula; the elimination rule allows us to take it away from a formula. We then see a proof as a set of manipulations of symbols from the formulas in the premises to the formula of the conclusion.[7] More philosophically, the elimination rules give us the strongest derivable formula, which is missing the symbol being eliminated. The introduction rules are what give the meaning of the connective; here, "meaning" really means the semantics. As we see, the semantics of intuitionist logic are contained in the rules, and they are quite different from those of classical logic. In intuitionist logic we have no need for semantic proofs *via* truth-tables or tree rules; we just have the introduction and elimination rules. We shall return to this.

Let us work through some more rules for the logical connectives. The rules for "or" are:

$$\frac{\Gamma : A}{\Gamma : A \vee B}$$

In words, "If from Γ we can prove A, then from Γ we can infer the disjunction A or B". Sometimes this rule is referred to as "weakening", sometimes as "or introduction". Again, "or" is commutative, so from a proof of B we can infer a proof of $A \vee B$. "Or elimination" is more elaborate.[8] The rule for "or elimination" is that if we have a disjunction $A \vee B$, and we can prove C from each of A and B separately, then C follows from $A \vee B$.

$$\frac{\Gamma : A \vee B \quad \Delta, A : C \quad \Theta, B : C}{\Gamma, \Delta, \Theta, A \vee B : C}$$

This bears a little discussion, especially for those brought up on a steady diet of Patrick Hurley or Irving Copi for their natural deduction. In the brazenly classical Hurley- or Copi-type systems, the closest we have to "or-elimination" is "disjunctive syllogism". Disjunctive syllogism says that if we have a disjunction, and the negation of one of the disjuncts, we can infer the other disjunct. In intuitionist logic, disjunctive syllogism is derivable but is not a primitive rule of inference. This is because when we move to a philosophical discussion about justifying the rules of inference, we do so with reference to the logical connectives. We then discuss these in terms of their introduction and elimination rules. If we have rules that combine more than one connective, then the philosophical discussion becomes much more muddied. This is because we are discussing the combination of two connectives, as opposed to one on its own, and these can only be justified philosophically by justifying the whole system of rules as a package. For the intuitionist, this sort of argument begs the question, since it is the whole system that is at issue. Instead, the intuitionist justifies each rule separately. So the intuitionist will derive rules such as disjunctive syllogism or De Morgan's laws.[9] If we do this, then we have a proof of them, and do not need to justify them philosophically. We only need to justify philosophically the first primitive rules and these should only involve one connective at a time. So for disjunction elimination we have to show that we can prove the desired conclusion from each disjunct separately (from each other). Then it does not matter which disjunct holds, or whether both hold; we have a proof of our desired conclusion. When we conclude C, we discharge A and B as separate assumptions. A, and separately B, were assumptions that were made for the sake of argument. Δ, A ∨ B A : C can be read as: "assume that from the disjunction A ∨ B, that A holds, then we can prove C". A ∨ B, B : C is read as "assume that from the disjunction A ∨ B, that B holds, then we can prove C". If we can get to C either way, then we no longer need to know which particular disjunct holds; we only need to know that the full disjunction holds, so we can do away with our assumptions. Discharged assumptions are sometimes symbolized with a bar over the letter for the formula. Using the bar notation in the or-elimination rule we would write the rule:

$$\frac{\Gamma : A \vee B \quad \Delta, \bar{A} : C \quad \Theta, \bar{B} : C}{\Gamma, \Delta, \Theta, A \vee B : C}$$

Consider implication. We shall use "→" for intuitionist implication. The elimination rule is essentially *modus ponens*.

$$\frac{\Gamma : A \quad \Delta : A \rightarrow B}{\Gamma, \Delta : B}$$

Implication introduction shows us the close rapport between implication and proof in intuitionist logic.

$$\frac{\Gamma, \bar{A} : B}{\Gamma : A \rightarrow B}$$

If from some formulas Γ, and an assumption A, we can prove B, then from the formulas Γ we can prove the conditional if A then B. We then discharge the assumption A. *Modus tollens* is worth mentioning as a derived rule. In intuitionist logic, *modus tollens* is carefully worded. From a conditional and the negation of the consequent, we can prove the *negation* of the antecedent. In symbols:

$$\frac{\Gamma : A \rightarrow B \quad \Delta : \neg B}{\Gamma, \Delta : \neg A}$$

What is careful about the wording, or the rule, is the use of the word "negation", as opposed to just saying "the opposite". By wording *modus tollens* through negation, some classical proofs will not go through, because of having to later use double negation elimination: for example, when A, in the rule, happens to be a negated formula. The astute reader will have noticed that in the *modus tollens* rule we are combining two connectives in the rule, which is something we objected to in principle when we dismissed the rule for disjunctive syllogism. This is a correct observation. *Modus tollens* is a derived rule: we can prove it from the other rules. We introduced it here to emphasize some points about negation.

Some general remarks are in order. There are different ways of presenting intuitionist logic. We can present it with a mixture of axioms and rules of inference. Presenting only rules emphasizes the notion that mathematics is a process, not a static discipline. In the presentation above there are only the rules of inference. It could be asked how anything is proved at all. Well, first one tends to be given premises, or hypotheses (assumptions for the sake of argument). So an argument is valid if its conclusion can be derived from the premises using the rules of inference mentioned above. Remember that there are three sorts of well-formed formula in logic: those that are always true, those that are always false, and those that are contingent on the interpretation. These are sometimes true and sometimes false, depending on the truth-value assignment to the proposition variables. The statement of validity is a conditional. It says that *if* the premises are true *then* the conclusion is also true. The "if–then" is subject to the rules of inference of the formal system governing the conditional.

To discuss proofs of conclusions from no premises (logical theorems), we should return to the notion of discharging assumptions. Or-elimination and implication-introduction are discharging rules. That is, one advances assumptions, for the sake of argument. These are discharged when the rule for the symbol is finally used for getting the desired conclusion. In the case of or-elimination, we assume one disjunct, prove the conclusion from it, then assume the other disjunct and prove the conclusion from it. We can then discharge both assumptions, since the disjunction alone is sufficient information to draw the conclusion. In the case of implication-introduction, temporarily, for the sake of the argument, we assume the antecedent of the implication. We then derive the consequent on the basis of the assumed antecedent, and then discharge the antecedent at the moment when we conclude the conditional: "if the antecedent, then the consequent". We can prove theorems using the discharging rules. Theorems are well-formed formulas that are provable from no premises. In classical logic, each theorem has a corresponding tautology.[10] In intuitionist logic we can prove $A \rightarrow A$ as a theorem. We can also prove $(A \wedge B) \rightarrow A$. Theorems will tend to be conditional statements, as they often are in classical logic. We can prove a theorem from no premises by making an assumption, which we later discharge.

For the most part, the rules for intuitionist logic should look very similar to the rules in classical logic. The big difference, besides the notation, has to do with negation. In some ways the rules are surprising. In intuitionist logic, it turns out that $A \vee \sim A$ is not a theorem. In other words, the law of excluded middle cannot be proved in the formal system. Similarly, the following proofs are classically valid, but cannot be carried out in intuitionist logic – they are not intuitionistically valid:

$$\frac{(A \mathbin{\&} B) \to C}{(A \to C) \vee (B \to C)} \qquad \frac{\mathord{\sim}(A \to B)}{A} \qquad \frac{(A \to B) \mathbin{\&} (C \to D)}{(A \to D) \vee (C \to B)}$$

The proofs are left for the reader to find. It is worth proving them in whichever classical system one is familiar with, and then noting where the proof would be blocked by the intuitionist.

As has been mentioned, there are other constructive formal systems, many of them more restrictive than intuitionism. For example, relevant logic systems block disjunctive syllogism.[11] The restrictions in other formal constructive systems have to do with implication, and how that is understood. As we also mentioned, the semantics for intuitionist logic are contained in the rules. The introduction rules are what give the meaning to the connectives. For now, let us leave the formal aspects of constructivism aside, and discuss some of the motivations for adopting a different logic to set a standard for reasoning in mathematics.

3. *Prima facie* motivations for constructivism

One quite effective *prima facie* motivation comes from considering the set-theoretic paradoxes. This motivation is only *prima facie* because it is not what is really driving the constructivists, but these considerations are enough to disturb any complacent acceptance of classical mathematics. Discussing the paradoxes gives us a reason to cast around for a solution, one of which is a rethinking of reasoning in mathematics.

Around the end of the nineteenth century and beginning of the twentieth century, a number of set-theoretic paradoxes surfaced, causing a crisis in the foundations of mathematics: in Chapter 2 we discussed the Burali-Forti paradox; in Chapter 3 we encountered the Russell paradox; and we briefly entertained a structuralist paradox in Chapter 4. This last paradox is very similar to Cantor's paradox, which we shall discuss here.

Cantor's paradox concerns the cardinal numbers. According to the theory of cardinal numbers, the powerset of a set has a cardinal number strictly greater than that of the original set (from which we "constructed" the powerset). Cantor's diagonal argument uses this fact. Consider the set of all cardinal numbers. What is its cardinality? Now take the powerset of this set. What is the cardinality of its powerset? The answer is that it is strictly greater than itself. Contradiction.

If we consider the Burali-Forti paradox, the Russell paradox and Cantor's paradox, then we have a good sample of the paradoxes that appeared in the early-twentieth century. The paradoxes made a mockery of contemporary logic and set theory, both of which were supposed to be foundations of mathematics. The Burali-Forti paradox and Cantor's paradox particularly address our conceptions of infinity, just as Zeno's paradoxes did over 2300 years earlier. These paradoxes partly motivate constructivism. One of the characteristics of constructivism is a distrust of the notion of actual infinity as it is developed in set theory. Not all the paradoxes address our notions of infinity; the Russell paradox does not. So, it is not only our notion of infinity that needs revising. After the emergence of the paradoxes, it appeared to many philosophers and mathematicians as though the whole of mathematics might be infected by these paradoxes and contradictions. Recall that from a contradiction anything follows (*ex falso quod libet*). So, a contradiction in the foundations of mathematics makes the whole of mathematics trivial, in the sense that any mathematical sentence is true, and false. In a contradictory theory with the symbols of arithmetic, 2 + 2 = 4, but also 2 + 2 = 5, and 5 = 39 and so on. Because of the threat of spread to other parts of mathematics, these paradoxes caused a crisis of confidence in the mathematical community. There were two sorts of reaction: keep all the mathematics we can and minimally tweak the axioms that give rise to paradox so that they no longer generate paradox (Zermelo's approach); or re-found mathematics on a firmer foundation by epistemically constraining truth. The former approach was adopted by most mathematicians.[12] The latter approach was much more radical, and this is what the intuitionists, led by Brouwer, proposed.[13]

4. Deeper motivations for constructivism

There are different degrees of commitment to the idea of revising mathematics by means of revising the logic that underlies the notion of mathematical proof. We can think of the new logic as setting a standard for excellence in proofs, or we can think of the logic as setting the minimal limit on what is an acceptable proof. If we think of the logic as normative, as setting a standard, we might or might not reach that standard in a particular proof. This does not imply that the purported proof is not a proof, only that it could be better. The normative constructivist counsels the mathematician to strive towards constructively acceptable proofs. This counsel is taken more or less seriously. Some mathematicians, who call themselves constructivist, merely acknowledge that the counsel is a good one, and flag all non-constructive moves in their proofs, more or less explicitly, with the idea that the proofs could either be rewritten in the future or, if we had to revise the theory because of a later paradox or contradiction, then we would know exactly which parts of which proofs to scrutinize. This sort of fence-sitting mathematician plays the game of caution in the following way. As a weak sort of constructivist, she does not believe that we have to start mathematics all over again. Rather, she believes that classical mathematics is for the most part fine. However, there are some problematic theories. A theory developed with only constructively acceptable proofs is guaranteed to be non-paradoxical, or is at least far less likely to be problematic, but the converse is not true. Not all non-problematic theories have to be constructive. Indeed, mathematicians of this inclination tend to think that the results we obtain using non-constructive means are usually true, and they point the way for a future revised proof that is constructively acceptable.

Why take this attitude? Constructive proofs are not always possible, and they require more work. We can get the results out more quickly classically. The results are usually good. We assume that we can, under the right circumstances of time and insight, go back and give a better proof later.

The problem with this practical fence-sitting attitude is that it gambles on the truth. This is loosely expressed in terms of probability: "probably" *most* classical theories are fine. What does this mean? If we probe, we find that it is difficult to say, since it is not clear that we can individuate mathematical theories in a relevant way to measure the probability, or even assert "most". With a little more charitable a reading, we can reinterpret what we mean by "probably" as drawing on the work of constructivists who do go back through the non-constructive proofs and give constructive proofs of the same theorems.[14] They are usually able to do this. The problem still remains, however, that from within this weak constructivist attitude there is no strong diagnosis as to why constructive proofs are better. After all, we can generate non-constructive proofs more quickly. Moreover, we should be aware that this talk of "more" is not really what is at issue at all. We do not simply want a large quantity of proofs. A computer can produce proofs at an alarming rate; moreover, it can generate constructive proofs just as fast as classical proofs. The problem is that most of these will be unimportant. In mathematics some proofs are more important than others; that is, they carry more weight, or they are more inspiring or more insightful.[15] Recognizing this, for the weak constructivist, is a separate (historical/psychological/sociological) issue. It remains that in the final analysis there will be proofs that are more significant than others; only time will tell. *Mutatis mutandis* for the argument that we favour constructive proofs over classical proofs (when the two are different) because constructive proofs give more information. This argument, too, is weak. Usually, this sort of argument refers to the difference between "purely existential proofs", which are classically acceptable, and "constructive existential proofs", which are constructively acceptable. Both existential proofs conclude that some object with a given property must exist. The difference is that the constructive proof has to generate an instance, or a witness, to the property. In other words, we need an object with the property in order to conclude that "there exists" an object with said property. In contrast, a classical purely existential proof usually runs through *reductio*: consider the idea that there is an object with said property, see that this leads to a contradiction, and conclude that there must be an object with the property. The classical proof gives us no means of finding an object with the property; it just tells us that there exists some such object. This is one piece of information, whereas the constructive proof gives us two pieces of information. On this simple numerical argument we should favour the constructive proof. However, think again. Recall that computers can generate proofs at an alarming rate, and just as many using constructive means as using classical means. Note that they will not necessarily generate the same conclusions, but we are just considering numbers of proofs, not types of conclusion. Now consider working human mathematicians generating proofs. We said earlier that many mathematicians find that they "can get results out faster" using classical proofs than constructive proofs, so now compare the mathematician allowed to use classical logic with the mathematician constrained by constructive logic. The classical mathematician will produce more conclusions, but the constructive mathematician will produce more information per proof, for certain sorts of proof, but fewer conclusions. Which mathematician produces more information? This is not clear. The very notion of "more information" is really a *non sequitur*. Ultimately, the fence-sitter is occupying an incoherent position. At best he has a superficial and unexamined view of quantity of results in mathematics or, at worst, he is gambling on classical logic's having a good handle on truth, allowing the mathematician's interests to be guided by the constraints of classical reasoning while at the same time acknowledging that this might be mistaken. In the end the normative or weak constructivist cannot mathematically, or philosophically, justify the preference for constructive proofs.

The philosophically more radical position is the revisionary position in constructivism. This is underpinned by well-defended philosophical considerations. Brouwer, considered to be the father of intuitionism, adopted some of Kant's ideas.

The strong, revisionary, prescriptive, constructivist philosophy is this: we should simply reject non-constructive proofs. Whole sets of results are dismissed as not part of mathematics proper. The reason for rejection is that classical mathematical thinking has gone astray, and has led us too far; it is no longer grounded in anything (independently of the classical mathematics itself).[16] The paradoxes are symptomatic of our having gone too far. Our pure, unchecked, thinking has led us into trouble. That is the diagnosis. So it is not that classical proofs are, for the most part, fine; rather, they are badly mistaken and incoherent. How do the strong constructivists come to think this? They think seriously about our knowledge of mathematics and the notion of mathematical truth. When we say that we have to epistemically constrain truth, we mean that, on pain of paradox, incoherence or irrationality, we cannot think of mathematical truth as independent of us. A truth is not such without a proof. Any objects discussed in a theorem have to be constructed, or displayed, and the proofs have to obey intuitionistically acceptable rules of inference. A metaphor that is often used is that we construct mathematics rather than discover mathematical truths. Mathematics should be seen as a process rather than as a body of truths.[17]

Brouwer discusses the process of mathematical construction, and is adamant that mathematics itself should not go beyond this process. In particular, the process of proving, or understanding, a part of mathematics has to be carried out in a step-wise manner. We have to understand and really follow each step in our mathematical constructions. The constructions and proofs have to accord with our basic mathematical intuition, not in the sense of Gödelian insight, but in the sense of each of us being able to follow each small obvious step. Mathematics, then, very much exists in our minds, and is not independent of us at all. If we take seriously Brouwer's way of thinking about mathematics, then we start to realize how fantastical the notion of an infinite set is. We cannot step-wise construct an infinite set. We cannot hold an infinite set in our minds at one time. The whole notion of an infinite set is quite incoherent. Recall the law of excluded middle. This says that if a formula is well-formed then it, or its negation, holds, where "holds" means not "is provable",[18] but rather "is true". "Is true" means that there is a model (i.e. a semantic interpretation)[19] for the sentence (or its negation), where "there is a model" just means that there exists a model, not that we know how to generate one or what one looks like. A model can be thought of as a domain that makes the sentence true. We might know of the existence of a model by means of a *reductio* proof; that is, the evidence we have for the model is that if there is no model then we get a contradiction. *Reductio* proofs are often called "indirect" proofs for this reason, and are not, in general, intuitionistically acceptable. In particular, they are not acceptable when they involve a double negation elimination step. Why does the intuitionist reject the package of double negation elimination, *reductio* proofs, law of excluded middle and infinite sets?[20]

As the modern proponents of intuitionism, we shall look at Michael Dummett and Neil Tennant's arguments. Following Dummett, we shall advance two arguments in favour of anti-realism: the acquisition argument and the manifestation argument. In *Elements of Intuitionism* (2000), Dummett brings these arguments to bear in favour of intuitionism. Neil Tennant deepens Dummett's arguments, and uses them in favour of a more restrictive logic, which he calls "minimal logic" in *Anti-Realism and Logic* (1987). We shall not pursue Tennant's arguments here all the way to minimal logic but, rather, use them to argue against classical logic, and we shall then default to intuitionist logic.

The acquisition argument and the manifestation argument arise from four considerations concerning language, meaning, communication and understanding. In particular, in this context, we are interested in language, meaning, communication and understanding in mathematics. The considerations bear outside the mathematical context as well, but this does not concern us.

> I. When we are mastering [a mathematical] language, and when we are exercising that mastery, all that is available to us for gleaning or conveying meaning is the overt, observable behaviour of fellow speakers.
> II. The meaning of any well-formed expression in our [mathematical] language depends in a rule-governed way on the meaning of its constituent simple expressions.
> III. When we have mastered a [mathematical] language, its sentences have a determinate meaning for us.
> IV. Our knowledge of those meanings can be displayed by an appropriate exercise of recognitional capacities shared by competent speakers [of the mathematical language]. (Tennant 1987: 3)

The acquisition argument concerns the first three considerations, in particular the first. In order to communicate we first have to listen, or read, to acquire a language. We have to grasp what others are trying to communicate to us. We are not born with a ready-made, particular mathematical language in our heads. We have to acquire the language from limited resources: the finite number of sentences uttered to us. We could not acquire a mathematical language without being exposed to some finite sample of sentences. We could not acquire the language without its being rule-governed. Otherwise, anything goes, and speakers are just making random noises. If there is no correct use of language then we cannot communicate. Implicit in this notion of the language being rule-governed is the idea that sentences (or well-formed formulas) have a determinate meaning. If they do not then we have no standards of correctness and, again, no communication, for, communication requires both a speaker and a listener. The speaker has to convey a message in a way that is graspable by the listener. The listener has to make the effort to understand what is being conveyed. For two-way communication to take place, both the construction of the language has to be piece by piece, and any reasoning, from one idea to another, has to be piece by piece. Otherwise, understanding is at best mysterious and inexplicable and, at worst, not happening at all. In other words, to acquire a language we have to observe speakers of the language speak (part I). The speakers have to be obeying some rules in their sound-making (part II). The rules have to be rules that we can follow. The sentences have to have some determinate meaning that we can grasp; otherwise we would not have acquired the language, but would just be making noises. If the noises have no determinate meaning, then we cannot be corrected in our noise-making (part III). This is the difference between communicating and making noises, which might, at best, have a psychological effect on listeners. The acquisition argument counsels us to make rules of inference explicit, and not require, or presuppose, great insight on the part of the student of mathematics.

The manifestation argument concerns being able to judge, of another person, that she really does understand what she says. That is, to belong to the community of communicators one has to manifest, to others, one's understanding of the language being used to communicate. We do this when we take a test in mathematics. We display our understanding to the teacher. In conversation, the manifestation is more subtle than a test, but nevertheless we can all think of times when we suspect that someone else does not really understand what he is saying. The person reveals this through his faulty manifestation of understanding, much to his embarrassment. For the manifestation argument, concentrate more on the fourth consideration above. We show our understanding through our behaviour, through the inferences we draw from a given piece of mathematics. This becomes quite serious in mathematics, where manifesting our understanding consists in proving, or showing that we can recognize or follow, a proof.

In mathematics we have nothing else to go on, save the proofs. Proofs are what allow us to demonstrate our understanding and knowledge of a mathematical concept. Moreover, not any proof will do. Listeners have to be able to follow the proof. This means that the steps have to be short, simple and indubitable. This explains why logic teachers are so exacting about proofs. It is through the proofs that the student manifests her understanding of the rules of logic. But why choose intuitionist logic over classical logic? Consider again the intuitionist rule for singe negation elimination. "Not-A" is concluded from "from A, we can prove a contradiction". This is because "not-A" is not just a negative fact; we have to be able to *show* why or how it is negative. Similarly, "not-not-A" is, intuitionistically derived from "from 'not-A' we can generate a contradiction". Notice the use of the word "generate". This highlights the idea that learning and acquiring knowledge in mathematics occurs through proofs and coming up with proofs; and following proofs is an activity. Does it follow from "from being able to generate a contradiction from A, we can generate a contradiction" that A? If we can prove a contradiction from A, then A is incorrect: it cannot be the case. Expanding again to the double negation, if we can prove a contradiction from the proof from A to a contradiction, then it does not follow that A is true, or that A holds, or that we are entitled to assert A. Trying to manifest our grasp of A by proving two contradictions is very convoluted, and not legitimate unless we have demonstrated our grasp of some separate background considerations that warrant our concluding that

A from the double negation of *A*. But, if we have these background assumptions, then we need to make them explicit in our proof.[21] So, classical double negation elimination is not intuitionistically acceptable as a law of logic. It only works provided there are other considerations present. The notion of manifestation in mathematical proof is one of carrying out demonstrations in small and easy to follow steps.

Armed with the acquisition and manifestation arguments, we can create arguments against the classical concept of infinite set. The classical conception involves manipulating infinite sets into the various transfinite cardinals and ordinals. With the acquisition and manifestation arguments, it becomes plain that we could not possibly manifest our capacity to take the powerset of an infinite set, since we cannot do this piece by piece. "Taking the powerset" is only a finite (and therefore acceptable) procedure if the original set is finite; it is not acceptable if the original set is infinite. It follows that there cannot be "different sizes of infinity". That is, Cantor's paradise is shunned. Cantor's diagonal proof is not intuitionistically acceptable. We can recreate similar arguments using the intuitionist requirements of acquisition and manifestation against classical *reductio* proofs, the law of excluded middle and the axiom of choice. (This is left for the reader to do.)

On a historical note, at a general level there are two attitudes towards the authorizing of proofs among the revisionary constructivists. Brouwer was adamant that it was not possible to formalize intuitionist logic; that is, it is not possible to decide in advance which proofs will be acceptable and which proofs will not. This is not very satisfactory since, in mathematics, we like to be very precise about what we are and are not allowed to do. In particular, we want to know which proofs are intuitionistically acceptable, and to know this, we need an explicitly developed logic. One of Brouwer's students, Arend Heyting, did propose an intuitionist logic. This has become quite well accepted by intuitionists. Dummett's intuitionist logic, presented earlier in this chapter, is equivalent to Heyting's logic. The logic is meant to constrain mathematical proofs, and can be used to tell us immediately, and relatively easily, whether a proof is intuitionistically acceptable. Current literature has moved on from intuitionism, and there are many constructivist positions. It is generally accepted now, by revisionary intuitionists, or constructivists, that there is some underlying logic. The intuitionists take this to be intuitionist logic. Other sorts of constructivist take other formal systems to prescribe the notion of mathematical proof.

To summarize, intuitionist logic is well accepted as a system of proof. Moreover, it is well defended as a guide to ensuring that, in mathematics, we properly manifest our understanding of mathematical theorems. However, ultimately, we should take Brouwer seriously too, and acknowledge that the formalized representation of intuitionist logic is ultimately revisable, in fact has been revised, and that each revision should be subjected to philosophical scrutiny.

5. The semantics of intuitionist logic: Dummett

For Dummett, semantics is completely articulated by the rules of inference of the system. He does not have separate truth-table proofs or semantic trees, although he shows how the rules of inference can be given tree-rule analogues. Dummett is philosophically austere because the natural deduction rules of inference are sufficient to talk about the meaning of formulas and connectives. He motivates intuitionism by thinking about proving as an activity. Moreover, to be a meaningful activity it has to be shared by a community, and has to be communicated. Dummett is an example of an intuitionist who will countenance bivalence, but denies that all well-formed formulas are meaningful. Notice that countenancing truth-values does not imply that Dummett thinks that we have to *define* the logical connectives using truth-tables. If a well-formed formula is not constructively provable using the rules of inference, then it is not meaningful, for we cannot know what its truth consists in, since we have no justification. We cannot know this because we lack a procedure for finding out. Similarly, we lack a procedure for sharing our knowledge with others. "Knowing" means "we can generate an intuitionistically acceptable proof". Talk of truth-values is parasitic on the rules of inference because truth is epistemically constrained by the rules. So while it might be convenient to talk of truth and falsity, such talk is strictly speaking redundant, and can be replaced by talk of proofs. Given this gloss on truth, it should be clear why Dummett accepts bivalence, but denies the law of excluded middle.

We have three notions interacting: being a well-formed formula, meaningfulness; and provability/truth. A well-formed formula is a candidate for meaning and provability. That is, if a formula is not well-formed, then it has no hope of being understood. For example, "*P*&&~" is a meaningless string of symbols. We could not possibly prove it, since there are no rules that will allow us to construct such a string of symbols. We can see this by inspection of the rules of inference for intuitionist logic. Compare this to a sentence in mathematics that is well-formed but still meaningless. Such a formula is one that we can recognize is well-formed, but for which we cannot see how we would prove it, even in principle. The rest of the well-formed sentences are proved, or provable, and therefore are meaningful, for, we can share the knowledge. We can show the proof. We can manifest our knowledge through proof and, therefore, we can share our knowledge through proof. Dummett insists that we cannot understand a formula we cannot prove or show leads to contradiction. We learn the meaning of sentences in a formal language through their proof, so proof is what gives meaning to a sentence. For this reason, only intuitionist proofs will count as proofs. Intuitionist proofs tell us how to go from one proof to the next without requiring special insight or powers. For the intuitionist, the semantics of a formal system is the proof procedure. All we need to understand the symbol "∧" are the inference rules governing it.

More specifically, repeating what we said earlier, Dummettian intuitionists will say that the introduction rules for a symbol give us the meaning of the symbol; the elimination rules tell us the strongest formula we can derive from a formula that has that symbol as the main connective in the original formula. For example, the or-introduction rule tells us the meaning of the symbol "∨". The question to ask is: does it really?

The interesting thing about this take on meaning is that, so far, we have nothing formal against which to judge these rules. They are the semantics, so the rules are what give us meaning. There is nothing to which they are responsible, such as truth-table definitions. The Dummettian might respond, however, that there is our natural-language understanding of "or", for example.

Why is this enough? There are three guiding notions to the justification of the intuitionist rules of logical inference: (i) we can only understand a finite number of things; (ii) language is compositional – we build our understanding from simple sentence to more complex sentences; and (iii) proofs are to be thought of as procedures for communication, not as static artefacts. Let us begin with the finiteness of our understanding. To do this we should add a note about intuitionistic quantification. There is an intuitionistic version of first-order logic. The universal quantifier (read "all") is understood as "we have a procedure for checking every" or "we have checked every". Universal quantification is fine over finite sets, but not so legitimate over infinite sets. In the following quotation, Dummett first discusses the classical mathematician's conception of meaning in mathematics, and then denies its plausibility:

> Since the theory of meaning underlying classical mathematics … consists in … an awareness of what has to be the case for it to be true, we must possess an understanding of [for example] quantification over an infinite domain which does not relate to our own restricted means of recognising as true, sentences formed by such quantification … The nub of the intuitionist critique of classical mathematics is the contention that we do not, and could not, have any such conception of mathematical truth; that we suppose ourselves to have it only by an illusion based upon a false analogy. (Dummett 2000: 258)

In other words, we cannot understand a sentence of the form "every natural number has the property F" since we do not have the means of surveying an infinite domain of objects. To think otherwise is simply delusional. At best, we might intuitionistically be able to say that "we have not yet found a number that lacks the property F". This is much more careful than the classical "all numbers have the property F", which we might prove classically.

Now we need an account of how it is that we do gain understanding of sentences. There are two components to compositionality. We learn the meaning of sentences by putting together words with which we are already familiar. This is how it is possible for us to understand a sentence we have never heard before. The "unit" of understanding is neither individual words (except when they constitute a whole sentence), nor the whole of the language, but rather the sentence. We understand whole sentences. We build our understanding from simple sentence to more complex sentences. Moreover, we understand simple sentences before we understand more complex sentences. Sentences can be arranged in a partial-order of complexity.[22] The logic we saw above is an intuitionist propositional logic. The units symbolized by the As and Bs above stand for formulas. The formulas are atomic or complex. If they are atomic, then they consist in a propositional variable, and these can have a determinate meaning. In particular, they do have a determinate meaning when they stand for a fact that can be proved, verified or shown to lead to a contradiction. If a formula is complex, then it is made up of propositional variables and logical connectives, so if the component propositions have a determinate meaning then the finite complex whole is either provable or its negation is provable.

The next element is the procedural aspect of proofs. Proofs have to give us a method for manifesting the meaning of a conclusion. The meaning of the conclusion depends on how we trace it back to some premises. The intuitionist rules of inference give us procedures for determining the meaning of the sentences in mathematics. Moreover, they are conducive to our necessarily finite understanding.

6. Problems with constructivism

The main complaint against the constructivist philosophy of mathematics is that it takes too much of mathematics away. When Brouwer suggested the refounding of mathematics on a constructive basis, tied to the intuition, he was met with fierce resistance.

> What Weyl and Brouwer do amounts in principle to following the erstwhile path of Kronecker: they seek to ground mathematics by throwing overboard all phenomena that make them uneasy [for example, the law of excluded middle] ... if we follow such reformers, we run the danger of losing a large number [infinite number] of our most valuable treasures. (Hilbert, quoted in Van Stigt 1998: 2)[23]

This is the main complaint against the revisionist constructivists. Douglas Bridges and Errett Bishop have, together and separately, taken this complaint as a challenge. They are working on recovering as much classical mathematics as possible using only constructive proofs. The task is daunting, but they are able to reconstruct quite a lot of mathematics. However, the valiant effort made by these constructivist mathematicians and philosophers somewhat misses the point: the debate is not about losing or giving up existent mathematical results. Instead, it is about those very results. For the intuitionist they are meaningless because they cannot be understood; there is nothing there to "recover".

If the constructivist is right, then by refounding mathematics we lose nothing except incoherence. For the Dummettian intuitionist, it is simply not rational to pursue mathematics using classical logic. Classical logic is not just dangerous, because it *could* lead to paradox, but is *incorrect*. So the quotation from Hilbert, and the feeling it echoes for many mathematicians, also misses the point. If the constructivist is correct, then we would not lose treasures; we would lose a few old rags.

The other problem the constructivist faces comes from the other direction: from other constructivists. There are different ideas as to what counts as an acceptable constructive logic underlying acceptable proofs. It is not clear, at this stage in our mathematical findings, whether one type of constructive logic is better at avoiding paradox than another, or sets a better standard of reasoning and of manifesting understanding. The debate rages, and in the meantime there are a number of new constructive formal systems being developed. There is too much choice; there are several different constructive logics and they disagree with each other. Each has some philosophical motivation behind it and each can, in part, be defended. The problem is to choose between them. Moreover, we do have to make a choice. What counts as proper, successful, coherent or good mathematics counts on it.

7. Summary

The important points to retain from this chapter are:
- Intuitionism is a particular type of constructivism.
- Constructivists are all anti-realists, and so strongly oppose realism in mathematics.
- The constructivist is concerned about the standards for correct reasoning and understanding in mathematics.
- Correct reasoning is important both for our acquiring mathematical knowledge and for our ability to manifest that knowledge so that we can communicate that knowledge.
- For the constructivist, the notions of truth and falsity are tied to knowledge. There are no verification-transcendent truths. Truth is epistemically constrained.

- The constructivist constraints on "correct reasoning" are restrictive with respect to the "results" obtained using classical proof techniques. This is a major complaint for the classical mathematician against the constructivist. But it is a weak complaint, since it misses the point.
- The more devastating criticism of constructivism comes from within: with having to choose between different constraints on "correct/acceptable" proofs.

Chapter 6

A *pot-pourri* of philosophies of mathematics

1. Introduction

Frege took exception to three philosophies of mathematics: empiricism, psychologism and formalism. His criticisms were so strong and influential that either little attention has been accorded to these philosophies, or they have been strongly modified or retrenched in answer to his criticisms. The criticisms can be found in the first half of Frege's *Grundlagen* (1980a) and in his correspondence, and they are well worth reading. In this chapter, we shall discuss these three philosophical positions as well as others that grew out of them. This will help us gain a better understanding of the positions we have discussed so far, since they will serve as a contrast.

Most of the sections of this chapter come in pairs, and should be read as such. Fictionalism (§3) can be seen as a rethinking of empiricism and naturalism (§2). Husserl's phenomenological approach (§5) is closely related to psychologism (§4). Hilbert (§7) is often thought of as a formalist (§6). The last two sections encourage a rethinking of the whole project of the philosophy of mathematics. The questions asked by a Meinongian philosophy of mathematics and by Imre Lakatos are different. Interestingly, all these philosophical ideas, from empiricism to Hilbert, are finitistic in some sense. Thus we return to our notion of potential infinity, and the view that mathematics is a process, not a static body of truths. So this chapter gives points of comparison with the subjects previously discussed and then goes on to give the reader a sense of the breadth of the philosophy of mathematics.

The first position we shall examine is that of empiricism. Roughly, this is the view that mathematics is simply an abstraction from empirical data. Empiricism is a type of anti-realism concerning mathematics in the sense that mathematics is parasitic on our experience of the physical world. So mathematics would not exist without the physical world to ground it. Furthermore, it is through sense experience of the physical world that we come to know mathematics. This implies that mathematical truths are neither eternal nor independent of the physical world. The champion of this position is J. S. Mill. Among current philosophers, this position is of interest to those who are wedded to a causal theory of knowledge, where this has the following stringent qualifications: (i) by "causal" we mean physically causes, where a physical theory of mechanics gives a canonical and implicit account of causation; and (ii) the only sort of knowledge that human beings can claim to have is causal in this sense. The result is that under a causal theory of knowledge we cannot literally (or directly) have knowledge of abstract entities. Either what seem to be abstract entities are not wholly abstract, or all of mathematics is a fiction. If we take the "not wholly abstract" disjunct seriously, we are led to naturalism. Central to naturalism are the so-called "indispensability arguments". These are arguments to the effect that some part of mathematics is indispensable to our best physical theory and therefore we ought to take that part of mathematics to be true. Note that this type of argument reverses the logicist hierarchy of knowledge; for the naturalist, physics, and our observation statements, are at the top of the hierarchy. They are the ultimate justifications for our theories, including our mathematical theories. If we take the other disjunct, and say that mathematics is a fiction, then we believe that there is not much difference between our knowing about the character Hamlet and our knowing about the natural numbers. The fictionalist believes that the sentences of mathematics are all literally false, in the same way as sentences about Hamlet are literally false. They can only be "true" when indexed to the context of the written work, in the case of "truths" about Hamlet that we get from the play; or traced back to a particular mathematical theory, in the case of mathematics. Outside these contexts the "facts" are false.

Quite different from the views above, we have formalism, which is seeing a revival. The revival, however, is largely implicit in the sense of not being widely published in philosophical circles. Formalism is the view that mathematics is a formal activity. What this entails is that, strictly speaking, mathematics has no meaning. All there is to mathematics is manipulation rules for the symbols. This resonates with intuitionist semantics. However, there are important differences between the intuitionist and the formalist. The formalist does not attribute meaning to mathematical expressions, whereas the intuitionist does. Moreover, formalists part company with the intuitionists when they insist that the rules of manipulation are somewhat arbitrary, in the sense that they need only to form a consistent system.[1] For the intuitionist, a formal system is ultimately responsible to our pre-formally expressed intuitions. Also opposed to formalism, we find the "classical" and/or realist view that the meaning of abstract symbols for logical connectives is given by the truth-table for the symbol. For a formal system, such as arithmetic, meaning is given by the numbers. Moreover, the semantic meaning relies on our insight into the true meaning and, ultimately, the formal representation is just that:

a representation of what we know to be true, independent of the representation. One of the keys to characterizing formalism lies in how the semantics of mathematics is analysed. For the formalist, there are no, or at least very few, external requirements on mathematics. In particular, applicability, appeal to intuition and truth are not required of mathematical systems. Importantly, for the formalist, mathematical sentences have no truth-value independent of the game in which they figure. The "game" analogy is taken quite seriously by formalists.

The current implicit revival of formalism comes from computer science. Computer scientists usually concede that computers do not understand what they do. They just perform according to a set of rules, but this strictly meaningless activity is mathematics, at least, this is mathematics to the computer scientist. Therefore, there is no reason to attribute meaning, in any deep experiential or emotional sense, to mathematics. The implicit formalist and computer science influence runs deep. Consider a certain style of writing mathematics or logic texts. In this style, the author is very careful to describe all calculations meticulously, so meticulously that we need no longer appeal to some intuitive idea. Older logic texts, and more classical logic texts, are not so explicit. Their authors take it for granted that all the readers "have logic" intuitively. So the latent mathematical abilities of the reader only need to be drawn out and given names. Long explanations, or explicit rules and instructions, are unnecessary.

Another implicit endorsement of formalism can be detected in recent developments in formal algorithmic learning theory. Some mathematicians, and some social scientists and economists, view human beings as "essentially" (analysable in terms of) algorithms. This view is the inverse of the idea that machines imitate or project our calculating abilities. Instead, human beings are essentially machines, superior at some computational skills than computers, and inferior at others. The point is that some sorts of computer scientist think of all of our mental processes as ultimately mathematically describable.

When the formalist takes this computer science inspiration, he sees our mental activity as essentially mechanical. This brings him close to psychologism. Psychologism (pronounced "psycho-low-gism" to distinguish it from "psy-cholo-gism", the more general attribution of anything at all to psychology) is the view that mathematical "objects" are mental constructs caused by, or supervening on, our brain configurations. Mathematics *should be explained* in terms of neuropsychology. Psychologism distances itself from empiricism. Mathematical truths are not grounded in the physical, outside us, but are grounded in mental activity. Thus, psychologism is anti-realist in the sense that mathematical truths are not eternal, and they are not independent of us. Whereas the psychologist might see the machine as an imperfect imitator of us, the formalist reverses this and sees us as an imperfect imitator of the machine. That is, for the psychologist we devised machines to calculate for us; they extend our abilities. The early champion of psychologism is Salomon Stricker,[2] but the position was also given some attention by Husserl. Arguably, Husserl also modified the view considerably, making his version of it an integral part of his phenomenology. Thus, we might call his more mature view the "phenomenological philosophy of mathematics". Stricker and Husserl were both heavily criticized and dismissed by Frege, but there has recently been a revival of Husserl's views. The revival both defends Husserl against Frege's criticisms of psychologism, and integrates Husserl's philosophy of mathematics into the rest of his philosophy.

When we discuss the philosophy of mathematics, we might ask different sorts of questions than the ones we have been asking so far concerning epistemology and ontology. Husserl is concerned with the phenomenology of mathematics; that is, with the nature of our (shared or objective) experience of mathematics. He is not so concerned with ontology, and his take on epistemology is subtly different from other philosophies of mathematics. In §9 we shall see another position that does not take ontology and epistemology to be the central questions in the philosophy of mathematics. Lakatos is interested in the development of mathematical thinking. He is interested not so much in the history (although he paid a lot of attention to it) as in the mechanisms for deep changes in mathematical thinking. We saw a little of this in Chapter 4, but whereas this is more of a diagnostic issue for the structuralist, it is a central issue for Lakatos. Interestingly, Lakatos believes that rather than really constructing mathematics like a building, with sets of truths carefully aligning with each other like bricks, we learn more from making mistakes in mathematics than by getting right answers. The refutations of purported proofs are what lead to the significant developments in mathematics. Yet another approach is to ask not "What is the essence of mathematics?", but rather, "What is the 'conceptual space' within which mathematical activity takes place?" To discuss this contextual space, the Meinongian philosophy of mathematics (§8), needs an underlying logic, and this has to be maximally permissive. So, as in Chapters 3 and 5, we see the deep relationship between logic and mathematics.

2. Empiricism and naturalism

"Naturalism" is the modern version of "empiricism". Empiricism was championed by J. S. Mill (1806–73), who wrote that assertions about numbers are assertions of observed, or physical, fact (1970: bk. II, esp. ch. vi, §2).[3] That is, for a diehard empiricist such as Mill, there is nothing beyond physical facts.

We begin with the very strong claim that *all* our knowledge is empirical. We learn to count by counting physical objects, such as pebbles. We learn to subtract by taking some pebbles away from our gathered pebbles. We observe that we have the same number of pebbles, even when we arrange them slightly differently. We can try to divide the number of pebbles in half by alternately placing them in two different piles, and then counting each pile to see if they have the same number of pebbles or not. Roughly, this pebble-counting story (Mill's analogy) accounts for arithmetic. What is important is that arithmetic begins with observations, and physical objects. Mathematics is an inductive science grounded in observation of the physical world. Geometry is presumably learned through engineering, through drawing and physical experimentation with the drawings.

The immediate problem with this position is that Mill, and other empiricists, are hard pressed to explain the existence of mathematics that goes beyond the observable. For example, we cannot straightforwardly claim that we directly see that the number of trees in the field is 0, for what we see are physical objects, not the absence of objects. We cannot touch, see, smell or taste absent objects. But recall Maddy's distinction between seeing and perceiving. Seeing is the purely physical activity that concerns the working of the eye and the reflection of light and perceiving includes a selection and interpretation of what it is we see. We cannot see 0 trees, but we can observe, or perceive *that* there are no trees. The empiricist now has to explain the difference between seeing and perceiving. One way of doing this is to follow Maddy. But we saw already that this approach suffers from implausibility. Recall that this was because Maddy's distinction between seeing and perceiving rests on our bringing our highly mathematically trained concepts to bear on what we see, in order to count as an act of mathematical perception. The problem with this, from the empiricist point of view is that this gets things wrong because the concepts have to emerge from the tangible physical world, not be imposed on the world from our mathematically trained minds. This is the whole point about empiricism. Mathematics is empirical. Perceiving has to be grounded in empirical facts, not be a mixture of empirical and conceptual facts. Similarly problematic is the observation that there are exactly 46,983 trees in a given delineated forest. We might count them, but with large numbers we could make mistakes, missing trees out, or counting some twice. Again, we might perceive that there are 46,983 trees. We do not directly see 46,983 trees. The numbers are too big to really see that many all at once. If we do see that many, say by looking over a valley on a clear day, then we cannot really say that this sight is different from the sight of 46,980 trees. We cannot perceptively distinguish these numbers. It is not our sight mechanism that allows us to tell the difference between the two numbers. We can only distinguish the numbers abstractly, for example, by saying that one is an even number, and the other is odd. The empiricist also has a hard time explaining how we might distinguish between observing that there are negative two pots in the sink from observing that there are negative three pots in the sink. All negative numbers *look* the same. Yet for the empiricist it is vital that mathematics be nothing above and beyond the empirical or physical world. So mathematics has to be explained in terms of the physical world, which is available to our senses.

There are two ways to go. The empiricist could bite the bullet and say that there are only a few finite numbers: those that we can observe directly, so "observation" does not reach beyond seeing. The rest of the numbers are a delusion, or are essentially interchangeable. There is no real difference between 46,983 and 46,980. Unfortunately, this sort of strict finitism is not very practical in our society, especially when we discount differences in negative numbers. It would be awkward to argue in a court of law that there is no difference between owing someone £3,000 and owing someone £30,000. In order to function in our society, the empiricist would have to learn a way of speaking that would allow him to function in such a way as not to give away his view that differences between very large numbers or negative numbers is, at best, approximate.

The other way to go is to try to be a little more sophisticated, and say that mathematics originates in sense experience, but much of mathematics is a *projection* based on the initial observations. What the empiricist concedes is some sort of induction, in the sense of generalization, from the finite observable empirical to the less observable. Then to account for numbers such as 0, negative numbers or large numbers, the observer generalizes on various rules such as addition, multiplication and negation. "Generalizing" just means going beyond the direct sighting. This is what allows us to do arithmetic in our heads, without physical objects in front of us. "Going beyond direct sighting" involves an imaginative leap, where we say something like "we can imagine counting 327 pebbles".

Apart from the difficulty of giving an empirical account of this projection (without reference to *a priori* mathematics), we could also ask about infinity. We could try to ground infinity in the physical by pointing out that the physical universe is infinite, or some aspect (time or space) of the physical universe is infinite. Some current cosmological theories assume the infinity of space and/or time, but it is not clear that they have to. After all, what cosmologists are trying to account for is the cosmos in time. The cosmos is really just the physical matter. There are a finite number of fundamental particles in this universe. The infinity of space or time is consistent with many cosmological theories, but is not necessary for these theories.

The empiricist has to offer a diagnosis as to what happens when we discuss, contemplate or reason about infinite numbers. There are two general strategies. One is to say that it is simply not "useful" to discuss infinite numbers; the other is to say that it is not "rational" to discuss such numbers. We saw the "not rational" strategy in Chapter 4. So, here, let us look at the first strategy. There is some confusion as to what counts as "useful", for it is usually a relative term: one thing is useful for something else. But be aware that whenever a philosopher cites usefulness one has to wonder whether there is any irony in what she says.[4] The empiricist cannot argue for infinite numbers in mathematics on the grounds of usefulness, for "useful" does not imply existence; there are "useful" fictions (we shall look at fictionalism in §3).

Returning to the charge of the uselessness of infinitary mathematics, we might say that for purposes of mere survival we probably need no more than the first five natural numbers. But this is not what the empiricist wants. The empiricist is not giving the philosophy of that part of mathematics that is essential to survival. Rather, she wants to give a philosophy of mathematics that is grounded in the physical world that we can observe. As long as we can use a number to count an object in the universe, then it is a number. Then we can ask: why is it that the empiricist chooses to limit herself to counting physical objects? Why not count abstract objects, such as concepts, or numbers themselves? Moreover, why does the empiricist think that she has the right to impose this choice on others? After all, mathematicians find it very useful to talk about infinite numbers.

Here is a diagnosis. The choice the empiricist makes has to do with trust. The empiricist is someone who, illusions notwithstanding, trusts her senses first, before any flights of reason (as opposed to the rationalist who trusts his reason above sense experience). This placement of trust seems to be a matter of personal taste or philosophical temperament. This is because the debate between rationalists and empiricists has not yet been resolved; and the issue is difficult, if not impossible, to resolve. This is indicated by the fact that the empiricist will (as will the rationalist too in his own way) start to beg the question against herself.

The conclusion is that there is a consistent empiricist position about mathematics. Mathematically, it is an agnostic position, metaphysically motivated by trust in observation statements. The negative side is that to the more rationalist-minded philosopher, empiricism and naturalism get the cart before the horse, for, in these philosophies what turns out to be good mathematics, and what turns out to be bad, is hostage to a choice of physical theory. This is backwards for the rationalist, because physical theory has to obey mathematics, and not the other way around. A physical theory that claims that $4 + 5 = 62$ is just not coherent; and yet the empiricist cannot reject that physical theory by appeal to mathematics, even in part. The choice of physical theory will require its own motivation, independent of mathematics. This sort of position will encounter fierce opposition from the rationalist.

3. Fictionalism

Fictionalism was developed by Hartry Field in his two books *Science Without Numbers* (1980) and *Realism, Mathematics and Modality* (1989). Fictionalism is the position that there is not a great deal of difference between discussing fictional objects, or fictional characters such as Moosbrugger,[5] and discussing mathematical objects such as the number 3. "Moosbrugger is a terrible, and simple-minded murderer" is literally false, because there is no Moosbrugger. Similarly, 3 < 7 is literally false, because 3 and 7 do not exist: they are literally fictions. A mathematical theory is like a work of fiction. We have to work within the constraints of the theory. The comparison between works of fiction and mathematical theories works surprisingly well. Not all fiction has been written. Similarly, there are many mathematical theories being developed. Mathematical theories are "finished" or self-standing when they have been axiomatized; all we then have to do is work out what follows from the axioms. Similarly, a work of fiction can be published, and then literary critics, or readers, are free to work out the implications of the fiction, or learn what lessons they can from it. We spend much time working out the implications of a work of fiction; see the many classes on English literature. Similarly, we spend much time working out the implications of certain mathematical theories; see the many classes in mathematics. Some works of fiction are given more attention than others, just like some theories in mathematics. Some works of fiction are more relevant to our lives than others. Similarly, some theories of mathematics are more easily applicable than others. There can be rigorous discussion about what follows from a work of fiction. In particular, if a work of fiction is found to be internally inconsistent, then we tend to reject it.

There are disanalogies too: with respect to mathematical theories, many mathematicians contribute to the development of one theory, whereas works of fiction tend to be written by one person. But this is just a coincidence. Works of fiction could be written by several people.[6] More importantly, we tend to think that while there might be vigorous disagreement about some interpretation of a work of fiction, there will be rigorous means of testing in the case of mathematics. That is, normative discussions in mathematics can be resolved, whereas often they cannot be resolved when discussing fiction. Sociologically, there is another important difference. If someone runs to a department of literature and exclaims "It is not literally true that 'Moosbrugger is a terrible, and simple-minded murderer'", he will be met with a shrug of the shoulders. Nothing surprising was exclaimed. In contrast, if someone runs to the mathematics department and exclaims "It is not literally true that '2 + 4 = 6'", she will be met with an incredulous stare. But how might the fictionalist position be defended?

The fictionalist contends that the difference between fiction and mathematics is one of degree, not one of type. While Clarisse might claim that

Moosbrugger is musical and, in the fiction, he does sing and dance briefly, it does not follow that he is really musical, not because he does not exist, but because of the lack of evidence in the fiction. The discussion cannot be rigorous because the terms are vague and ambiguous. What it means to "be musical" might be different for Clarisse than for an arbitrary reader. We cannot further interview Clarisse. Or we might also disagree as to what counts as evidence that someone is musical; his once dancing and singing in his prison cell is not enough to claim that Moosbrugger is musical. The issue cannot strictly be resolved in discussing fiction.

In contrast to literary fiction, ambiguity and vagueness are maximally expunged from mathematics. This is what allows us to have rigorous discussions and proofs. The purposes of writing fiction and writing up a mathematical theory might be different, but the fictional nature of the two is the same, according to the fictionalist. The differences are of degree, not of nature.

Mathematical "truth" is relativized to a theory, just as a fictional "truth" is relative to a work of fiction.

> [T]he fictionalist can say that the sense in which "2 + 2 = 4" is true is pretty much the same as the sense in which "Oliver Twist lived in London" is true: the latter is true only in the sense that it is true *according to a certain well-known story*, and the former is true only in that it is true *according to standard mathematics*. (Field 1989: 3)

The question now is: should we believe "standard mathematics"? More carefully, do the objects of mathematics exist, and is standard mathematics a theory about these? If, along with the fictionalist, we only trust our sense data as giving evidence for something existing, we are inclined to answer yes to the first question, and no to the second, for mathematics is a sort of convenient fiction. The standard mathematics is "standard" in virtue of its being readily applied to the physical world. However, this is not enough, says the fictionalist, to warrant the ontological claim that mathematical entities exist. Evoking the indispensability argument, mathematics is essential to physics, and it is that essential part of mathematics that is standard.

The naturalist is someone who argues that whatever part of mathematics is indispensable to our best physical theory is true, since physics is true. So, the entities posited by the mathematical theory exist on a par with the entities posited by the physical theory. That is, the mathematics cannot be divorced from scientific explanations. It is not as though we have one part that is the physical part, and one part that is the mathematical part. The "two" are completely enmeshed. If well run, the indispensability argument is enough to convince the philosopher who favours sense-perception data, over pure logic and rationality as a source of knowledge, to accept "standard" mathematics as true.

The fictionalist resists the conclusion drawn by the naturalist, and says that the mathematical "entities" are nonexistent because not physical. The argument runs as follows. In *Science Without Numbers*, Field shows us how to do some science (mechanics) without literally appealing to numbers. In other words, Field shows that the entities of arithmetic, namely numbers, are dispensable with respect to our physical theory of mechanics. The idea is that we should be able to extend the project to explain more of physics, chemistry and biology. In so far as we can do that, we do not really need numbers to do science; numbers have been extricated from science. The ontology underlying standard mathematics has been shown to be separable from physics, so the ontology of standard mathematics is not indispensable to our scientific theory. This is supposed to undercut the indispensability argument for believing in mathematical entities. Standard mathematics is not literally true.

There is a glitch. Philosophers are sceptical about how far we can extend the programme initiated by Field. In particular, they notice that the notions of space and time, so in particular all the space-time points, are not eradicated from Field's version of mechanics. To do mechanics, we have to be able to distinguish space-time points from each other, and once we start to do this, we have enough mathematical entities, namely, all the members of the continuum, to re-establish the ontology of standard mathematics as indispensable. This is not a final criticism. The fictionalist can take it as a challenge: to see if we can extend the programme to really reproduce the results of the whole of physics without ever making appeal to a set of mathematical objects, such as numbers. Also, it is worth noting that in order to undercut the indispensability argument, it might be enough just to show that what is indispensable to physics is not some *unique* part of mathematics. Rather, we might be able to show that there are several different, equivalent, ways of re-expressing the results of physics, each appealing to a different mathematical theory, with different mathematical objects (real numbers, natural numbers, lines and points on a graph, etc.). The different theories would be equivalent with respect to the central virtues of the physical theory, such as predictive power, explanatory power and fit with other theories. What this would show is that there is no privileged mathematics that is indispensable to physics, but rather that there is a disjunction of mathematical theories, one of which is indispensable: either arithmetic, or analysis, or geometry, or topology, or set theory and so on. We can choose which one on the basis of efficiency, familiarity, fit with other theories, or other considerations that are not purely mathematical. The indispensability argument is then substantially weakened, for it then says that in so far as we are committed to the entities in physics we also have to be committed to some mathematical entities, but it is not at all clear which ones.

This "disjunctive version" of the indispensability argument is not enough to vindicate fictionalism, for we cannot conclude that therefore *no* mathematical entities literally exist, so all mathematics is literally false. The arch-enemy of the fictionalist, the realist about mathematical entities, could simply retort that it is true that there are a number of different sets of mathematical entities, each a candidate for indispensability to physics. However, one of them is the privileged class of mathematical objects, we just do not know which one on the basis of physics alone. Our physical theory is not discriminating enough to tell us which one. We simply have to look at other evidence to narrow the pool of possibilities. At best, the fictionalist project is too incomplete to be well supported. At worst, it is not possible to complete it.

The other major problem that the fictionalist faces is to account for the rigour of argument in mathematics. Many mathematicians and philosophers instinctively feel that the rigour of argument is significantly different from that of arguments concerning fictional characters. It is not just a matter of degree, as we suggested above, but a difference in type. For this reason, the fictionalist argument from analogy with fiction is not convincing to many mathematicians or philosophers. The dispute concerns whether the arguments in mathematics really are of a different type than those concerning fiction. Here we reach a stalemate, or challenge, since it is not clear what would count as showing that the rigour we are used to in mathematics is "of a different type" than the rigour we see in arguments about fiction.

4. Psychologism

Psychologism is similar to fictionalism in the sense that the objects of mathematics, such as numbers, are mental rather than physical constructs. The psychologist in interested in the word "mental": in "mental construct". The fictionalist is interested in the metaphysical implications of the claim that mathematical objects are mental constructs. In contrast, the psychologist reduces mathematics to activity in the brain. Mathematics is no more than brain activity. Thus, $2 + 8 = 10$ is no more than a series of mental computations that, with training, some of us find easier to do than others. Stricker was one of the first to propose psychologism.[7] His version is rather crude and easy to ridicule. However, psychologism takes an interesting direction when we look at Husserl's phenomenological approach to the foundations of mathematics. In some places Husserl seems to be a psychologist. However, it is also clear that his position is not entirely psychologist, for it is intimately related to his development of phenomenology.

According to Frege, Stricker claimed that mathematics is in us and is dependent on psychology. "Stricker … calls our ideas of number motor phenomena and makes them dependent on muscular sensations" (Frege 1980a: v). Frege goes on to lampoon this position and makes two important points.

> [N]o mathematician can recognise his numbers in such stuff or knows what on earth to make of such a proposition. An arithmetic founded on muscular sensations would certainly turn out sensational enough, but also every bit as vague as its foundation. No, sensations are absolutely no concern of arithmetic. No more are mental pictures, formed from the amalgamated traces of earlier sense impressions. All these phases of consciousness are characteristically fluctuating and indefinite, in strong contrast to the definiteness and fixity of the concepts and objects of mathematics.
> (*Ibid.*: v–vi)

As Frege describes it, the psychologist position is easy to dismiss. However, let us take a closer look at it, while addressing his general points against the psychologist. We can update the reference to "muscular sensations" in the first quotation by replacing it with "neurons firing", and similarly update "motor-phenomena" by replacing it with "neurological phenomena". The first point Frege makes is that the mathematician cannot recognize his mathematics in a neurological description; the second is that psychology cannot account for the "definiteness and fixity" of mathematics.

On the first point, psychologism, like fictionalism and Hellman's structuralism, is an eliminativist philosophy of mathematics. According to the psychologist, we should be able to re-express mathematical sentences in terms of neurological descriptions. This is a radical idea: that we should be able, once we have done enough neuroscience, to rewrite mathematical textbooks with neurological descriptions instead of equations written in mathematical language. The textbooks would look rather different than they do today. Frege's point is that it is not at all clear that the researcher in mathematics would find such a book illuminating at all, with respect to mathematics, although, as Frege says, it would be sensational enough, in its own right. There are different degrees of eliminativism. The most radical version would say that we ought to replace all our talk of mathematics with talk of neuroscience, so rewrite the textbooks. The less radical version is to accept that, while in principle we could do this, it is not practical to do so. Nevertheless, the less radical version continues, the truths of mathematics are not based on a fiction, or on real entities. Rather, mathematical truths are based on how our brains are constructed. So we could rewrite the textbooks, although for practical reasons we will not insist on this; but if one wants to know the deeper truth of the matter, then one had better start reading neuroscience. The radical version is difficult to accept now, given our present understanding, and at best looks like something we could only do in the distant future, if at all. The less radical psychologist position also involves a future projection: we still have to show that we can map the equation $6 + 7 = 13$ to some neurons firing, and $6 + 18 = 24$ to some other neurons firing, and at this stage in brain research this is still a distant hope.

On the second point made by Frege, a sceptic can resist the reduction from mathematics to neuroscience. Frege's point is that the psychologist could never hope to explain why arithmetic is this way and not another way, for, Frege might argue, our mental make-up is accidental with respect to arithmetic. Our brains do not determine mathematics. Our brains enable us to discover mathematics. It is necessary for us to have brains to do this, but that is not all there is to mathematics. Our knowledge of mathematics might depend on our having brains, but the brains do not *explain* mathematics. According to Frege, arithmetic could not depend on how our brains are constructed, because if that were the case and our brains were constructed differently, our arithmetic would be different. In other words, psychologism makes mathematics implausibly subjective, because it makes it dependent on brain configuration. For Frege, the objectivity of mathematics rests not on a biological or neurological fact, but on a conceptual epistemological fact, lying outside us. Consider our practice of mathematics. If, in a classroom, a child learning to add were to say to the teacher "I feel/sense/compute that 2 + 8 really equals 12", it would be ridiculous for the teacher to reply "Oh, well, you just have a different arithmetic, due to your personal neurological make-up. I'll send you to the surgeon to fix the problem". The aberrant adding is not fixed with surgery; it is fixed by aligning the concepts. But, the psychologist has to defend the idea that, given enough neuroscience, it would not be inappropriate to send the aberrant child to the surgeon, rather than go over the meanings of the terms in the equation. Again, this sort of science-fiction mathematics is unrecognizable to us now.

The psychologist might reply that modern neuroscience has the hope of drawing a fairly accurate map of the brain in the relatively near future. So the charge of "subjectivity" that Frege levels at psychologism is no longer apposite. It might well be the case that there is a definite and precise rapport between certain mathematical operations and particular brain activities. In other words, there might be a reduction of mathematics to brain activity, and there might be some sort of explanation of mathematics in terms of brain activity. Finding the mechanisms, and being able to surgically, or chemically, alter them, is only a question of time.

However, one has to ask, as Frege did when he says that "the mathematician cannot recognise his numbers in such stuff", why one would be interested in this reduction, and what sort of explanation is being given. This sort of explanation might explain how, or why, it is that a particular person is having difficulty with learning division, or why it is that we need so much practice in order to grasp certain mathematical operations. This sort of explanation might help with tolerance in teaching mathematics, but it is not the sort of explanation that will help with developing further mathematics. This is why the mathematician "cannot recognise his numbers in such stuff". This is also why this reduction or explanation is important to the psychologist, neurologist, cognitive scientist, or teacher, but not to the mathematician.

However, the psychologist might still maintain that brain activity is essentially what mathematics is. The reduction indicates the origin; and the origin gives us the essence and the limitations of the subject. Maybe some things are inconceivable (cannot be configured in the brain), and therefore will never be a part of mathematics. If this is right, then the reduction would have philosophical importance, if not mathematical importance. However, Frege has a reply to this too.

> Never let us take the description of the origin of an idea for a definition, or an account of the mental and physical conditions on which we become conscious of a proposition for a proof of it. A proposition may be thought, and again it may be true; let us never confuse these two things. (1980a: vi)

A philosophy of mathematics is supposed to justify mathematics, not account for our thinking it at all. Psychologism might account for our interest in, or activity involving mathematics, but it cannot account for the truth of mathematics.

On careful examination, Frege's rebuke is not so strong. While it may be appropriate to draw this distinction sometimes, it is not clear that it is entirely appropriate in this case. To claim that the distinction is appropriate, here, one has to appeal to something like the independence of mathematics to the thinker of the mathematics; and this is exactly what is at issue. The psychologist cannot divorce the thinker from the mathematics. The realist can and has to. Thus the two are talking at cross purposes.

Moreover, psychologism also implies that consensus over, say, basic arithmetic, depends only on the brain's configuration. Once the psychologist argues this, then she also has to *account for* the consensus over arithmetic truths. The psychologist could say one of three things: (i) it is miraculous, or coincidental (which is not a good move in philosophy); (ii) there are evolutionary reasons that explain the consensus; (iii) in deconstructionist mode, we might venture that there actually is little consensus, we just think there is.

The evolutionary account will not do. It is better suited to supporting realism. Consider the fact that different groups of people, isolated from each other, still develop the same arithmetic, albeit to different stages. In so far as this is the case, then that development of the same arithmetic rests on the underlying reality of arithmetic as being in the world, since it is evolutionarily advantageous for us to cotton on to the real arithmetic, and not something else. On this account, we select for the real arithmetic, and then the arithmetic is real independent of us. It is in the world: our evolution helps us to grasp it, but does not create it.

The last tack for the psychologist is more radical, and essentially involves a deconstruction of mathematical practice. The psychologist could argue as follows. We delude ourselves when we say that there is complete consensus concerning mathematics. There is no fundamental consensus. Instead, consensus is a social construct. There is enormous normative pressure in the classroom to get the mathematics right. We learn to conform to the thinking the teacher projects, for sociological and psychological reasons. These are not mathematical reasons. As students, we feel the psychological pressure and we respond by suppressing our first instincts to add in our own way, and reinforce the teacher's way in order to conform.

This story, too, is not very plausible. First, many people are instinctively rebellious, and no one thinks of developing odd mathematics as a political ploy, which it would be if the deconstructionist psychologist's story were a good one. Secondly, the classroom-pressure idea only makes sense in a modern setting, where we have the right sort of classrooms. Mathematics is very ancient. It was developed independently in different parts of the world. The teaching of mathematics was carried out in different ways, and none allowed differences of opinion on basic equations to go unchecked.

In conclusion to this section, it is probably most devastating to the psychologist position to point out that this reduction is in no way helpful to the mathematician *qua* mathematician, as opposed to a teacher of mathematics, or neuroscientist, to whom findings about the relationship between brain activity and mathematical calculations in human beings might be helpful. Even if the neuroscientist were to say that we can develop mathematics by stimulating certain parts of the brain in the right order, or something like that, this is still not satisfying for the mathematician, for she is not interested in brain activity. She is interested in mathematics as a subject that is treated as incidental to particular brain activities. The means of stimulating mathematical thought should not be confused with the subject of mathematical thought. The elimination of mathematics in favour of psychology is too distorting to be of interest as a philosophy.

5. Husserl

Frege accused Husserl of being a psychologist, and some of the remarks Husserl makes would certainly support the accusation, especially in some passages of his early writings. However, some philosophers today argue that it is too simplistic to view Husserl's writings on mathematics as psychologistic.[8] In fact, he was deeply distrustful of psychologism, even before Frege's accusations.[9] Instead, these philosophers argue that it is better to think of Husserl's philosophy of mathematics in the context of his phenomenology.

Thus, Husserl is better characterized as giving a phenomenological philosophy of mathematics.

What does this mean? The phenomenologist is someone who takes our experience of the world to be the most fundamental area of enquiry. The point of philosophy is to enquire about the nature of our experience of the world, and mathematics in our case. Experience involves both sense experience and mental activity. Central to the mental activity is intentionality. Intentionality bears on, and modifies, propositions. Propositions are facts. Grammatically, propositions are announced by the word "that" in English: "that Elizabeth had her hair done", "that the cat is on the mat", "that 2 + 5 = 7". Different sorts of intentional attitudes (modifiers/operators) include belief, knowledge, fears, wishes and so on. Intentionality accounts for our focusing our attention on an object. It also gives information about the attitude, or *type* of attention, we are bringing to bear on an object. The phenomenological approach to the philosophy of mathematics consists in reporting on experience, not on a personal level, but on a general and impersonal level. The purpose is to understand the very nature of experience itself, as a general, intentional, not personal, phenomenon. The following explanation of phenomenology should help.

> I begin ... by inviting you to engage in a very simple exercise ... This exercise involves little more than continuing to do what you are doing right now, which at least includes looking at the words printed on the page in this book. ... That you are looking at the words on this page, that you are reading, means, among other things, that you are engaged in the act of seeing, or, to be a bit fancier but perhaps no less awkward, that you are currently having or enjoying visual experience. Now, suppose you are asked to describe *what* you see. In response, you may note such things as the page before you, along with the words and letters, and perhaps also the shape of the page, the shape and colour of the letters. You may even read aloud the words that are occupying you the moment the request is entered. You may also, if you are being especially careful and attentive, say something about the background that forms a field on which the page appears. ... [C]onsider a slightly different request. Instead of being asked to describe what you see, the "objects" of your visual experience, suppose you were asked to describe your *seeing* of the objects. Here, you are being asked to shift your attention away from the things you see to your visual experience of these things, and here you may find the request a little less straightforward ...
>
> I happen to wear glasses. If I were to take them off while looking at the page of the book held at the usual half-arm's length away, the letters, words and page would, as I might put it, become blurry ... That there are descriptions that apply to visual experience without necessarily applying

to the objects of that experience helps to make vivid the distinction we are trying to delineate between what we see and our seeing of it. To concentrate on the latter, to focus one's attention not so much on what one experiences out there in the world but on one's experience of the world, is to take the first step in the practice of phenomenology. The word "phenomenology" means "the study of phenomena", where the notion of a phenomenon coincides, roughly, with the notion of experience. Thus, to attend to experience rather than what is experienced is to attend to the phenomena …

… Phenomenology … invites us to stay with what I am calling here "the experience itself", to concentrate on its character and structure rather than whatever it is that might underlie it or be causally responsible for it. (Cerbone 2006: 2–3)

Husserl's idea is to provide a "science of consciousness". We study consciousness in a systematic way, not as an object as we do in neuroscience, but as "noetic" experiences: experiences that lead us to knowledge, understanding and experience of the world. What Husserl is interested in is the relationship between knowledge and experience: how experience can teach us anything at all. Our consciousness is what allows us to know and understand, and make sense of the world around us. This knowing, understanding and making sense of the world is subjected to analysis by Husserl. There is a certain structure to experience, and there are pre-conditions to experience, since experience is intentional. That is, we must have intentionality, which presupposes that we have brought our attention to bear on an "object of experience" in order to have an experience of an object. Our conscious experience has to have structure; we usually have some battery of concepts that we can bring to bear on our experience, since these will shape our experience. If we now consider mathematics and logic, the analysis becomes quite interesting, for it is our experience of abstract objects that we analyse: how we come to know and understand these. Further, we want to understand the interplay between our experience and knowledge of the abstract with our knowledge and understanding of the concrete. Husserl is deeply impressed by the rigour and objectivity of mathematics. Moreover, mathematical objects, in the sense of objects of study, are "ideal objects". They are not physical, and we do not experience them through our sense perceptions. Nevertheless, they are entirely objective. We cannot change them at will.

We can understand better if we contrast Husserl's phenomenological approach to mathematics to both empiricism and psychologism. The empiricist believes that our knowledge exclusively stems from experience of the physical world, and cannot reach beyond this experience. Here experience is very much thought of as sense experience. The empiricist does not enquire into the nature of experience *per se*. This is taken for granted, and as primitive, so it cannot be further analysed. The empiricist is interested in grounding, or justifying, our knowledge by tracing it back to sense experience. The empiricist trusts his senses. In fact, knowledge can only be gained through sense experience. My seeing three trees before me is good evidence that there are three trees before me. In contrast, the phenomenologist is interested in the more abstract questions of what it is like to be seeing three trees, and what it is to count three of them, and this has to do with our intentional stance towards the objects of attention, not with our sense experience of the objects. The phenomenologist is not interested in gathering empirical facts, and ascertaining how sure we can be of these facts but, rather, in reflecting on the activity of mathematical fact-gathering as one type of experience that we have. Moreover, Husserl is quite aware that work with mathematics reaches well beyond the physical world. In fact, he agrees with Frege that mathematics itself comes before the physical world. The *application* of mathematics to the physical world is a different sort of activity or experience from that of thinking of pure or ideal mathematics.

Let us turn to the contrast between the psychologist and the phenomenologist. The psychologist reduces mathematical activity to brain activity. The forming of mathematical concepts is reducible to physical changes in the brain. The phenomenologist is interested not so much in particular mathematical experiences, but in the experience of mathematics quite generally. The exploration of this will not, for the phenomenologist, degenerate into a discussion of neurons firing or other brain activity; nor will it degenerate into a discussion about certain brain types having more or less propensity for mental calculation. Rather, the phenomenologist is interested in what happens conceptually when we form a concept of a number, not with respect to our physical brains, but with respect to how this changes our perception of the world, how it might influence our further experience, or how it is that we come up with a judgement involving mathematical concepts. For example, the phenomenologist might ask how it is that we come to say that there are three trees, or that there are no lemons. More interestingly, the phenomenologist will enquire after, say, an axiom of infinity. How do we justify this? What is our experience of the justification, and how does our experience of abstract objects justify the axiom?

Note that Husserl is not interested only in easy and elementary mathematics. He studied a lot of mathematics and was, for 15 years, a colleague and friend of Cantor (Hill & Haddock 2000: xi). So he was well versed in very abstract mathematics. What does it mean for us to be conscious of a new (to us) mathematical object? We have to have some intention towards the object. Maybe this is provoked by curiosity, or a prompting from a teacher. Our intention has to be precise enough for us to recognize the object when we grasp or apprehend it. We have to be able to distinguish that object from another, that is, say, distinguish the infinite set of natural numbers from the infinite set of even numbers. These are purely mathematical experiences. When we learn some new mathematics we learn to discern new mathematical objects, or relations between mathematical objects. We witnessed this in the discussion of the actual infinite in Chapter 1. Which mathematical sentences are true is told to us by mathematicians. Which mathematical objects exist is also told to us by mathematicians or by our grasping the truth or by our generating a proof. What is more interesting is what the bounds of conceivability are. Cantor pushed these bounds further than anyone before him. To discover what is conceivable, we have to enquire into the concept. We make a phenomenological analysis of the concept. The phenomenological analysis of mathematics is quite different from the analysis of empirical truths, for mathematical objects of attention, or objects of study, are quite definite in the sense of being well-defined; and the same mathematical object of attention is studied by different mathematicians. This last point is important, and draws out another meaning of "objective" when applied to mathematics. For Husserl, there is no doubt that the number π that was studied by the Pythagoreans is the same as the number π being calculated by modern computers. The objects of attention of mathematics transcend time, space, culture and particular personal experiences. To emphasize this point, Husserl calls mathematical objects "ideal objects". Recall the distinction between objectivity in ontology and objectivity in truth-value. Husserl's objectivity is different. We might call it "objectivity in phenomenology". It is objectivity *in the way the object* (of our attention) *is presented to us*. Mathematical facts are hard immovable facts.

The objections to Husserl's phenomenological philosophy of mathematics will be quite deep, for they will involve questions about the success of the phenomenological approach, and to judge success in this case we also have to say something about the very reasons for doing philosophy. These issues have been raised in previous chapters, but in a more superficial way, for in previous chapters the differences in approach *to philosophy* were not so very great. Husserl has a truly different approach.

To fix on the area of discussion, it is useful to compare Husserl's phenomenological approach to mathematics with Frege's logicism. Husserl's correspondence with Frege is revealing because both are interested in the same questions: what a mathematical object is, and whether, or to what extent, logic is foundational to mathematics. The difference between them lies in what they will accept as answers to these questions or, more precisely, what the presuppositions are to any answers. Husserl agrees with Frege that logic plays a special role with respect to the rest of mathematics. For Husserl, logic is fundamental. "Logic" for Husserl means "fundamental in the structure of experience". That is, Husserl recognizes that logic will inform our mathematical enquiry. Logical moves are phenomenologically more basic than other mathematical moves. However, the phenomenology of our mathematical enquiry is what we are seeking to analyse, not some Fregean "ultimate justification". For Frege, logic is fundamental in the sense of offering an ultimate justification. In particular, Frege was afraid that Husserl's approach involved too much psychology. Frege was adamant that his ultimate justifications were free of any psychological connotation. Frege's conception of knowledge is divorced from psychology because Frege was afraid that by letting psychological considerations slip in to an analysis of knowledge this would make mathematics look personal, and mathematical truths look relative, and subjective. Mathematics would then be true for a person, or for his psychology. We can have knowledge of psychology. However, for Frege, this knowledge will not afford us insight into anything but psychological matters.

Arguably, Frege mistook Husserl's phenomenological approach for a psychologistic approach.[10] The mistake is easy, since Husserl does not believe that the philosophical investigation of mathematics concerns only the logical justification of a realm of truths that is independent of us. The immediate supposition, made by Frege, is that Husserl must be interested in the psychology of mathematics. Frege misses Husserl's point about intentionality, which sits between logical justification and psychology but cannot be reduced to either one. Husserl agrees with Frege that, as philosophers, we should avoid any sense of the personal, or the subjective, in analysing mathematics. The misunderstanding arises because Frege cannot see the absence of personal considerations or psychology in Husserl's phenomenological approach. Frege gives a false dichotomy. Either we give a psychological account of mathematics, or we give an ultimate justification (based on logic). For Frege, only the latter guarantees the objectivity of mathematics. Husserl does not pin the objectivity of mathematics to a hierarchy of justification, or of knowledge. Husserl takes an intermediate approach: he talks of "ideal objects", which are abstract objects. But this is not enough for Frege. In contrast to Frege, for Husserl, logic does not provide a justification for anything. This is not Husserl's interest. There is no reason to justify anything in the way that Frege does. Instead, we enquire into what being conscious of mathematics consists in. The disagreement between Husserl and Frege concerns what counts as answer to a philosophical question.

Again, "objectivity" is very different in the two cases. For Frege, mathematical truths are objective in the many senses we saw in Chapter 3. Frege felt compelled to show, or prove, that mathematical truths are objective. For Husserl, the "objectivity" of mathematics is evident. What is not evident is how it is that we come to grasp these objective truths and work with them to develop and discover more mathematics. This was the real focus for Husserl.

Husserl and other philosophers of mathematics differ over what the philosopher can tell us about mathematics, and what counts as a good answer to the central questions. Interestingly, the central questions often sound the same. In order to disagree with Husserl, we have to disagree that the central point of philosophy is to offer a theory of consciousness, or that the way to develop such a theory is to engage in phenomenological analysis.

6. Formalism

Compare fictionalism to formalism. The fictionalist thesis is that mathematical sentences are all literally false because they fail to refer to anything. Mathematical sentences can only be true within the context of the mathematical theory, but the theory is literally false. In contrast, one of the main characteristics of formalism is the view that mathematical sentences are literally meaningless. For the formalist, mathematical sentences are the results of manipulations according to rules, and so cannot have a meaning. This is because the manipulations are mechanical and thoughtless. There is, therefore, no content or meaning. In *Reason's Nearest Kin*, Michael Potter puts it very well:

> The fact that quantifier-free elementary arithmetic[11] reduces to the mechanical application of a finite number of rules allows us to decouple it from its meaning: there is an obvious sense in which this sort of simple arithmetic does not require *thought* at all.[12] It is natural, therefore, to wonder whether we can obtain an account of arithmetic that focuses entirely on the signs and abandons any attempt to argue that arithmetical propositions express thoughts about a subject matter distinct from the signs occurring in them. (2000: 10)

The idea is that we should be able to extend this natural thought, about quantifier-free elementary arithmetic to the rest of mathematics. Mathematics is neither mental (psychologism) nor based on the physical world (empiricism). Instead, mathematics is simply a series of mechanical procedures. Mathematics consists in symbol manipulation. An upshot of this is that the meaning, in so far as there is any, is not contained in the semantics, as realists conceive. Rather, the meaning of the mathematical symbols is derivative, not literal and, strictly speaking, dispensable. Meaning, in so far as there is any, is contained in the rules for manipulating the symbols. Since, to most philosophers, this is a degenerate notion of meaning, let us refer to it as "manipulation-meaning". Manipulation-meaning falls short of regular notions of meaning. In particular, consider that, for the formalist, sentences in mathematics do not get a truth-value, except as part of a game internal to mathematics, but then it is not a real truth-value. The truth-values T and

F could be replaced by 0 and 1, or any other symbol we choose. Attributing truth-value is just another set of mechanical manipulations. The manipulation-meaning is dependent on the game being played, so all we can really say is that a sentence is in accordance with the rules, or it is not.

This philosophical idea about manipulation-meaning is partly born out in some logic textbooks, which introduce natural deduction before they introduce the truth-tables.[13] The idea is that the manipulation rules are sufficient to give the meaning of the connectives. As we saw, some constructivists believe that the meaning of the logical connectives is entirely revealed by the manipulation rules for those connectives. Nevertheless, for the constructivist, the sentences still have meaning or content, for we can be called on to justify a manipulation rule. The truth-tables are an alternative, realist, way of understanding the connectives. But each of these positions is different from the formalist stance, which says that the sentences of mathematics are literally meaningless. We just have moves in a game.

The textbooks that begin by introducing rules for manipulation only partly bear out the philosophical position of formalism because they do, typically, later introduce truth-tables.[14] A purely formalist textbook would never mention the truth-tables, except as a philosophical, or historical, note. Or it would introduce the truth-tables as a mechanical game we can play, where any notions of truth and falsity would be suppressed because they are meaningless symbols in a game of making "truth-tables". Typically, we suppress these notions by making T = 1 and F = 0. It is no coincidence that the computer scientists prefer 1 and 0 over T and F. Another feature of a formalist-type text in mathematics is that it will be exceedingly explicit about what to do with each symbol. There are no informal discussions about what a symbol "means intuitively". These are replaced by manipulation rules. Similarly, there need not be any suggestive names or shapes for the symbols. The formalist idea can be detected through the whole of mathematics, beyond introductory logic.

What does the mathematician do, according to the formalist? She devises, experiments with, or finds fault with, different sets of manipulation rules for symbols. One can imagine the mathematician beginning with an existing formal system, modifying some of the rules and seeing whether new theorems can be proved, or whether there are theorems provable in the new system that are not provable in the old system. She will have failed in creating a new system if the new system is provably equivalent to the old, or if an inconsistency is derivable from the new system.

An advantage of the formalist philosophy of mathematics is that it is conceptually very free. Pure mathematics is not responsible to anything external to mathematics; the constraints on mathematical activity are purely internal. They consist in a demand for consistency, complete rigour and explicitness about the rules (Curry 1963: 11).[15] This is conceptually liberating because the mathematician can treat mathematics as a game of manipulation, and really explore what happens when one adds a new operator to a language, or a new rule of manipulation. The mathematician does not have to justify doing this in terms of wanting to prove a theorem, or in terms of responding to some pre-mathematical (platonic or realist) intuition about what there is, or in terms of the importance of the new system to some application. The "game" analogy is taken very seriously. In fact, this is exactly where complaints about formalism are usually aimed, for the freedom is a double-edged sword. There are not enough constraints, so there is no way to select a direction for trying to develop mathematics. There is no part of mathematics that is more important than another. We just spin the mathematics in the void; there is nothing to ground it conceptually. We would be better off letting computers take over, since they are faster and less prone to error. We have no mathematical, objective or debatable justification for playing the game in the first place, for we are not discovering truths. We cannot judge mathematical activity, except to say that it is successful in some applications. Applicability is seen as an *ad hoc* issue: to do with applied mathematics, not pure mathematics.

This is an oversimplification because it is not always easy to separate the pure interests from the application interests. The formalist has to be careful to say that what is important, or interesting, or applicable in mathematics is not a matter for *mathematics* to settle. That does not mean that these matters cannot be settled but just that they have to be settled from outside. For example, the proof of Fermat's last theorem has historical importance. It might even have a romantic, or sentimental, importance. But the formalist recognizes all these measures of importance as lying outside mathematics itself. The result, and the proof, might be helpful for other parts of mathematics. But even this is, ultimately, a feature of what mathematics we have developed and which problems we *want* to solve; and which problems we want to solve has to be analysed, according to the formalist, in non-mathematical terms. Choice of a particular mathematical pursuit is primitive and extraneous to mathematics. That is, a mathematician simply says that she is interested in this branch of mathematics, and cannot be asked for further justification.

What the formalist will stringently avoid saying is that we "discover" truths when we prove results in mathematics. The formalist will also avoid saying that we can be certain that there are more results to be proved, for the already existing results are not couched in a body of truths ready for our discovery. The formalist position is definitely anti-realist. This, in turn, suggests that the formalist might take a stance on the issue of infinite numbers. The argument in favour of the formalist being a closet finitist runs as follows. The formalist believes that mathematics is simply symbol manipulation. We, as finite beings, can only perform a finite number of manipulations on a finite number of symbols even if we extend our powers with computers. This entails that there are only a finite number of proofs that we shall ever generate. The finite number of symbols and manipulations also entails that we shall only ever have a finite catalogue of results. This does not entail that we cannot manipulate symbols that the platonist thinks of as symbolizing infinite numbers. So for the formalist, there is nothing wrong with writing "$\aleph_4 + \aleph_6 = \aleph_6$"; this sentence is just as meaningless as "$21 \times 1 = 21$". Each is a string of symbols. Each is a legitimate equation in a game of addition or multiplication, respectively. But neither sentence is literally true; it only follows the rules set out by the game, or the manipulation rules applied to the axioms of the theory.

However, we now have a problem. We have to ask how it is that adding infinite numbers does not, *prima facie*, obey the same rules as adding finite numbers. Following rules for addition in finite numbers, we would think that $\aleph_4 + \aleph_6$ should really equal $\aleph_{10}$. But no existing mathematical theory endorses the equation $\aleph_4 + \aleph_6 = \aleph_{10}$. This is because $\aleph_6$ is so much bigger than $\aleph_4$, it just absorbs $\aleph_4$ without noticing. We have not added anything of significance to $\aleph_6$. All addition of infinite cardinals is like this. Adding an infinite cardinal to any other cardinal is equal to the greater of the two, or the same, if they are the same. How might the formalist explain this? The formalist is reluctant to say that what might have motivated the choice about how to understand addition with infinite numbers was some sense of what the infinite numbers are. For the formalist, this way of speaking is ultimately misguided. There is nothing that the infinite numbers *are*; there are just various symbols that we manipulate. Strictly *mathematically*, we made a choice about addition of the infinite cardinal numbers that is consistent with the other manipulation rules concerning infinite cardinal numbers. We did not reveal some deep truth about infinity. We simply adopt conventions in mathematics that are consistent with previously set-out rules, and follow our rules through. The real answer, for the formalist, is that we generate a contradictory theory if we endorse "regular" arithmetical rules when adding with infinite cardinal numbers. If we generate a contradiction, then we have a trivial mathematical system, in the classical sense of "anything follows". That is, every well-formed formula both holds and its negation holds. If we have a trivial system, in this sense, then we are not playing a game, says the formalist. We do not have a good set of rules. Notice that we did not say that a trivial system is meaningless for, strictly speaking, all mathematical formulas are meaningless. This is not a way of distinguishing trivial from consistent mathematical theories. Why the need for consistency? It is a primitive constraint on the notion of a "good game".

Unfortunately for the formalist, this is deeply unconvincing to most mathematicians. They feel constrained to add infinite cardinal numbers in a certain way because of how they understand infinite cardinal numbers. If their understanding of cardinal numbers is not a mathematical matter, then it is not clear what it is. The realist mathematician can justify the requirement for consistency in a theory as a minimal requirement, among others such as truth. The formalist thinks of consistency as a "primitive" constraint: one for which the justification cannot be forthcoming.

There are philosophically weaker replies to this. These mainly involve a retrenching of positions. The arguments are ones we saw in Chapter 5, since the issues we raised there are not dissimilar to those that the realist will raise against the constructivist. According to the formalist, the greatest resistance to formalism is not made on mathematical grounds but, rather, on emotional grounds. Since these are not mathematical, they need not be addressed.

Stronger replies by the formalist to the emotional and phenomenological complaint involve reference to computers and modern developments in theory of computation. Arguably, computers and theory of computation carry out the formalist programme, for computers extend the thoughtlessness metaphor. What a computer does is manipulate; it does not think. Computation theory shows us that it is possible to calculate solutions well beyond quantifier-free elementary arithmetic. The formalist seems to characterize the mathematician as a lesser sort of computer, with no legitimate mathematical or philosophical grounds for why she is carrying out the manipulations. Similarly, computers do not have to be convinced to make an effort to make a calculation; they just carry out their program. Computers are physically less demanding than we are, and they do not get tired as easily; of course, a computer might still run out of electricity, parts of computers physically wear out and so on. Nevertheless, they are faster and more accurate than we are. In fact, computers are able to do quite sophisticated calculations. The formalist position assumes greater strength when we consider the great sophistication of these computers, for they are also able to imitate us quite well, including making mistakes if we insist on them. Computers can be designed to pass the Turing test.

The Turing test is about imitation. The test is set up as follows. We put a computer in one room, and a person in another. A second person, who does not know which is in which room, is allowed to type questions to the occupants of the rooms. The second person is trying to guess in which room we have a computer, and in which room we have a person. He asks questions that he hopes will reveal which is which. Since current computers are quite good at imitation we can program them to make human-type mistakes, and give human-type answers; we can even program them to imitate impatience, or other emotions. We cannot always program computers successfully to imitate human beings, but we are getting ever closer. This is a major part of the artificial intelligence business. The Turing test, if successful (that is, if the second person is unable to tell the difference between the person and the computer), is supposed to tell us that there is no *real* difference. Philosophers then worry about what *real* means in this context. Now add the formalist twist: restrict the "players" to a mathematician and a computer doing mathematics. The third person asks mathematical questions. We can always build the required human time lag and human mistakes into the computer. The point of the Turing test adapted to the formalist position is to tell the difference between the results produced by a mathematician and those produced by a computer. Knowing that we can slow a computer down, and have it "make mistakes", the computer might well pass the Turing test and be mistaken for a mathematician. Once the computer passes the Turing test, and is mistaken for a person, it is an easy step from this perfect imitation to saying that mathematicians, or any people, are equivalent to computers with respect to mathematical content (i.e. there is none). Ability and speed is a different issue. The formalist is vindicated by the Turing test since we can give a complete algorithmic account of our mathematical activities: thanks to the sophistication of current computers and current algorithmic learning theory.

It is worth pausing now to compare formalism to psychologism. Interestingly, the psychologist can also run the mathematical version of the Turing test. But he will conclude something quite different from the scenario where the computer passes the test. This is where the psychologist and the formalist part company. Whereas the formalist is delighted to reduce mathematics to computer activity, since this is clearly simply symbol manipulation, to the psychologist, what is important is that by developing computers we have extended *our* capacity to calculate. Computers have been constructed to imitate human beings, not the other way round. As such, they extend *our* powers, which originated in our psychology. Which way the reduction goes is what distinguishes the psychologist from the formalist. The formalist reduces humans, and especially mathematics to computation. The psychologist points out that in designing a computer to pass the Turing test, computers are imitating human beings, and so the essential characteristics we are trying to draw out are psychological characteristics, or brain activity, not pure manipulation. Since we have these two ways of reading the results of a successful Turing test, the test is not decisive in deciding between the two positions. But it is an interesting thought experiment, which does force us to ask questions about whether computers are like people, or if people are like computers. Either way, the realist will be quite unhappy since brain activity or computer activity do not capture mathematics because they miss out what is important in mathematics: the objective truth, or meaning, of mathematical discoveries.

Returning to the finitism question, computers are finite machines, and can only carry out finite calculations. Manipulations, are, by the nature of the manipulators, necessarily finite. Nevertheless, we can certainly manipulate symbols that the realist thinks of as referring to infinite totalities. But the manipulations themselves are finite. In this sense, the formalist is a "closet finitist". There is only a finite amount of mathematics. The symbols thought of by the realist as referring to infinite totalities, do not so refer. The notion of infinity as a mathematical object, or mathematical notion is meaningless. We shall explore this issue further in the next section, when we discuss Hilbert.

Returning to our emotional and phenomenological complaint above, the bottom line is that the formalist's terse characterization of mathematical activity is resisted by many mathematicians. Mathematicians tend to have realist leanings, and feel that they are doing something important. Furthermore, the importance of mathematical activity does not lie simply in the application of mathematics to subjects outside mathematics. There is a sense of discovering truths independent of us. This is the draw enjoyed by the realist, or platonist, positions. These positions carry intuitive sway, but they are not easily defensible, as we saw in Chapter 2.

Hilbert is unhappily characterized as a formalist, for he, like many mathematicians today, is interested in what lies beyond what is manipulable and (anachronistically speaking) computable, and is reluctant to make the philosophical move of dismissing what lies beyond. He recovers this infinite part of mathematics, and calls this the "ideal realm". Note that he does not mean this in Husserl's sense (of abstract). Hilbert's "ideal realm" is the mathematics beyond finitist mathematics: the infinite ordinals and cardinals. Because Hilbert is interested both in the finitistic and the ideal realm, he is like many current mathematicians. Nevertheless, when talking of the realm of mathematics that falls short of the ideals – the finitistic realm – he sounds a lot like a formalist.

7. Hilbert

David Hilbert is famously remembered for a talk he gave at a large international mathematics conference hosted at the Sorbonne in 1900 in which he listed 23 problems for the mathematical community to solve over the next century. Mathematicians took him seriously and the set of problems was widely distributed, most of them being solved by the end of the twentieth century. There is a theme to the problems; they come from Hilbert's vision about what was important in mathematics. He not only contributed substantially to setting the agenda for mathematics in the twentieth century, he also made major contributions to pure mathematics.

In part, Hilbert's vision arose from a plan he had for mathematics. Hilbert wanted to secure the foundations of mathematics against contradiction by giving finite and rigorous procedures for working in mathematics. This was Hilbert's formalist and anti-realist side. But he is not best classified as a formalist, finitist or anti-realist, for he believed that classical mathematics, including the mathematics of the infinite, was all good mathematics. In particular, Hilbert was fascinated and enchanted by Cantor's development of the transfinite numbers. He was not interested in seeing anti-realist philosophers of mathematics, such as Brouwer, "drive us out of the paradise that Cantor has created for us" (quoted in Shapiro 2000: 159).[16] So Hilbert has two sides: the formalist (anti-realist) side and the realist (classical) side. Confusingly for the language as it is used by philosophers, Hilbert talks of "real" propositions and "ideal" propositions, where these are philosophically reversed. Hilbert's "real" propositions are what the anti-realists are interested in, and Hilbert's "ideal" propositions, are more easily associated with realism and classical logic. Hilbert's formalist tendencies were motivated by concern about the paradoxes, but he remains realist because he is quite convinced that classical mathematics is all good. Hilbert wanted to prove this by deriving ideal mathematics from the secure, paradox-free, finite mathematics. Securing all of mathematics this way is known as Hilbert's programme.

Hilbert placed great emphasis on the axiomatization of mathematical theories, and on giving rigorous deductions of theorems in those theories. The purpose of this was to prove a point against Brouwer, and wean mathematics away from intuition. The more explicit and mechanical the derivation of a theorem, the less it relies on our feel or intuition. In fact, the calculator illustrates this very well. We are able to have a calculator churn out a number of mathematical results without having any inclination to attribute to it mathematical prowess or mathematical intuition. This is possible thanks to our ability to make the instructions for calculation explicit and determinate.

Like formalists, Hilbert thought that rigorous axiomatization together with explicit rules of inference demonstrate that intuition is strictly redundant with respect to the subject of mathematics. Hilbert brings this approach to geometry. The idea was to make the axioms of geometry quite explicit, and the rules of inference quite determinate. Once this is accomplished, Hilbert demonstrates that, *pace* Kant and Frege, geometry does not rely on an intuition of space.[17] Extending the approach to other branches of mathematics is Hilbert's programme.

Say it *is* possible to carry out Hilbert's programme for the whole of mathematics. Then it would seem that any formal system with sufficiently explicit rules constitutes mathematics. Again, as we saw with formalism, this becomes a very liberal attitude towards mathematics. Mathematics is a series of games, and there are an infinite number of possible games to play. One is not better than another, if we are judging by strictly mathematical criteria. In fact, the only criterion for completely ruling out a game is inconsistency. No consistent formal system is deemed "crazy", at least not on mathematical grounds.

Hilbert was well aware of the possible excesses of such a liberal attitude. He did propose some more constraining guidelines on the choices of mathematical games, and he packed this into his notion of "finiteness". These were not absolute constraints, but more of a general guide to choice. We have to be careful about what we mean by Hilbert's constraints. The strongest one is consistency. As a comparative aside, note that strictly speaking, for a realist, demonstrating consistency is unnecessary (Hallett 1991: 2), for our insight into the truths of mathematics is what guarantees consistency. We perceive mathematical truths through intuition or whatever. Consistency is not something we need to prove; instead, it is a precondition of thought, especially about a mathematical notion. Of course, the period around the end of the nineteenth century into the beginning of the twentieth century was one where paradoxes were surfacing at an alarming rate. The paradoxes made it plain that our intuitions are not wholly reliable guides to consistency in mathematics. So for the realist proofs of consistency are checks on the reliability of our intuitions, or perceptions (however these are explained); they are not a check on the mathematics. In contrast, for the formalist consistency is a criterion for good mathematics, not for good intuition.

For Hilbert, consistency had to be demonstrated by giving a model. In his geometry Hilbert showed that the formal geometrical system was consistent with the arithmetical characterization of the real numbers. This is a relative consistency proof, that is, it shows that Hilbert's geometry is consistent if and only if the arithmetical characterization of the real numbers is consistent. What is important in Hilbert is that the real numbers are not then treated as intuitively obvious. That is, the proof is not an absolute guarantee of consistency. Rather, the real numbers are a model for the geometry. The model is not treated as a semantic entity, in the sense of giving meaning, but as part of a syntactic process of translation from geometry to arithmetic.[18] The model does not ground, or give meaning to, the geometry. So Hilbert understands the notion of model as an ethereal "end of a syntactic process", as with geometry, or as a move in a meta-game, sanctioning the original game (of geometry).

There is now a very great danger that the theory of real numbers is shown to be inconsistent. If that were to happen, then it would show that geometry and analysis are trivial, in the sense that every mathematical sentence and its opposite hold in each system. We then no longer even have a proper game going, because all moves are fine. This possibility is a lingering problem in mathematics. To this day we have no absolute proof of consistency of the major parts of mathematics. In fact, in 1932 Gödel showed us that if a mathematical theory of a certain minimum complexity is consistent, then an absolute proof of its consistency is impossible. Only a relative consistency proof can be given. If the mathematical theory is not so complex, then we can give an absolute consistency proof.

Hilbert's major contributions predate Gödel's results, so he would not have known about the impossibility of absolute proofs of consistency. However, he still sensed some danger for he felt that there should be a second constraint on mathematical activity: he chose "finiteness in procedure". We have to be careful. Hilbert was very fond of Cantor's infinite numbers and he was not interested in curtailing manipulations of these. But Hilbert did urge the distinction between finite mathematics and ideal (infinite) mathematics.

Hilbert distinguishes three types of proposition: "strictly finitary", "finitary but general" and "ideal". The first two were considered to be subdivisions of "real" propositions. Strictly finitary propositions are ones we find in quantifier-free elementary arithmetic: "37849 − 7893 = 30956" is an example of such a proposition. We can check whether it is true or false by running a finite check over particular finite numbers. We can determine that the equation is false because each number in the equation is finite and there are no quantifiers or variables. "Finitary but general" propositions are ones with variables, each of which can be substituted with a finite number: for example, "$5 + x = x + 5$". We want to say that this is true, since this is just an instance of the commutativity of addition. Now, x is assumed to range over finite numbers, so there is an implicit bounded quantifier: "for all x, where x ranges over the (finite) natural numbers". We do not bother to make the quantifier explicit, since it is understood. Moreover, it is understood as a substitutional quantifier; that is, $5 + x = x + 5$ is true whenever we plug in the same finite number for the x on each side of the identity symbol. We know that this is true, but we do not prove this by checking every finite number. This would require an infinite proof. Nevertheless, because the implicit quantifier is bounded by the finite natural numbers (it reads "substitute any finite number for x") the equation is considered to be real, and so finitistically acceptable. Ideal propositions are all the rest. These have either quantifiers ranging over numbers that are not finite – for example, "for all cardinal numbers", "for all ordinals", "for all real numbers" – or unbounded quantifiers, where it is left free to choose a domain, and the domain could be of any cardinality. For example, Frege's basic law V: $\forall F \forall G((\text{Ext}F = \text{Ext}G) \leftrightarrow \forall x(Fx \Leftrightarrow Gx))$ is an ideal expression. The quantifiers range over predicates, and so over subsets of a domain. Moreover, since the proposition/expression/well-formed formula is meant as a law of logic, and logic is universally applicable, we should be able to bring any domain we like for the quantifiers to range over. "$\forall x$" is not generally restricted (bound) to the finite natural numbers. Unfortunately for Hilbert, most interesting mathematics concerns ideal propositions, and not real ones.

Hilbert hoped that all of mathematics could be shown to be a conservative extension of "finitary" arithmetic. That is, he hoped to show that the "finitary but general" propositions and the ideal propositions were just shorthand for strictly finite propositions. Since we can, in principle, capture all of quantifier-free elementary arithmetic by means of a wholly mechanical finite procedure, so we should be able to capture other parts of mathematics too. One way to do this is to concentrate on our finite proofs and the use of signs. If we write "$\aleph_0$," we write a finite sign. In a realist moment we might interpret this to refer to a size of set, and that size is infinite, but it is still represented using a finite sign. Similarly, all of our equations are written using finite signs. We can assign a unique number to every symbol in a language and thereby number our equations. This is called "Gödel numbering". Gödel numbering gives us a method for demonstrating sort "finiteness". When we are thinking this way about mathematics, we return to the contentless, and meaningless manipulation of symbols. Unfortunately, the device of Gödel numbering, and this finitist way of thinking about mathematics will not, in fact, get us very far in Hilbert's programme. Gödel shows us that not all of mathematics is a conservative extension of finitary arithmetic. Hilbert's programme, understood in the sense of demonstrating consistency of a formal system finitistically, cannot be carried out. On the other hand, we can salvage something by turning this into a limitative result, telling us that Hilbert was quite right to distinguish between the conservative finitist part of mathematics and the ideal part of mathematics. Hilbert's programme can be given new breath by expressing it as an investigation into where the border lies between real and ideal mathematics. In a sense, this is what the mathematical discipline of proof theory is doing.

Because Hilbert is trying to show that mathematics is finitary, in some sense, this implies that he took arithmetic to be primitive, or beyond question. Depending on how we understand this, we then distance Hilbert from the formalists, since he is essentially grounding mathematics in arithmetic, which he sees as true, as opposed to "yet another game". This leads to some philosophical problems in Hilbert, since he then has to say something about what makes arithmetic (which grounds the very notion of finite procedure) true. In other words, Hilbert has to tell us on what basis he favours the finite part of mathematics over the ideal. His answer has to do with concrete physical signs, such as strokes: "3" is really better represented as "|||" – three strokes. We can then calculate and discuss concrete signs, and not abstract ideas. Nevertheless, there is an uneasy tension in Hilbert's notions of meaning, symbols and manipulation. For "|||" represents something; it is not *prima facie* meaningless. If we turn now to Meinongian philosophy of mathematics, we see what happens when we loosen Hilbert's criteria.

8. Meinongian philosophy of mathematics

Meinongian philosophy of mathematics is named after the Austrian philosopher Alexius Meinong (1853–1920). We shall begin with a word of warning. The very mention of Meinong's name sets off alarm bells in the minds of many philosophers, because Meinong's writings were heavily criticized and ridiculed by Russell in 1905 and 1907. Most philosophers took Russell's criticisms to be decisive, and simply did not bother reading, or trying to defend, Meinong's theories. Current proponents of Meinong claim that Russell misread Meinong, and they therefore feel entitled to return to his writings. Meinong's views are not wholly unattractive. We shall discuss Meinong's views in general, and then turn to the philosophy of mathematics, which makes use of his views.

Meinong had an original theory of ontology. He was interested in accommodating our talk of fictional objects: of exotic objects that could not possibly exist, such as a mountain made of gold, and even contradictory objects such as round squares. To accommodate this talk, Meinong distinguishes between "existing objects" and "subsisting objects": objects that have properties but do not exist. An object might have "sein" (being) and it might have "sosein" (how it is, i.e. a description or characterization).[19] "Being an object" does not imply "existence"; it only implies "having properties". Every object has properties. Some objects exist, and some do not. When we are discussing fictional objects, we attribute properties to them but not existence. For example, Hamlet has the property of "being prince of Denmark". Hamlet does not have the property of "being an old woman", at least not on a straightforward reading of the play by Shakespeare. Similarly, the "mountain made of gold" is an object, but it does not exist. It is an object in virtue of it's being the bearer of properties.

Richard Routley takes Meinong seriously and adapts Meinong's insight to mathematics. Routley maintains that in the actual world only concrete objects exist. Thus, anything abstract – relations, numbers, measurements, ideas, properties and so on – do not exist in the actual world. They do, however, have properties, since even properties have properties. Properties of properties are second-order in the sense of being one level of abstraction up: rather than being properties of objects, they are properties of the properties of objects. "Is a colour" is an example of a second-order property, and "blue" is an example of a first-order property. "Hat" is an object. First-order properties characterize objects. In particular, properties characterize objects that exist (in the actual world) and possible objects, which exist in a possible world. We discussed modality a little when we discussed Hellman's structuralism.

Recall that the idea behind "possible worlds" is to make sense of talk of possibilities, and possible objects (technically "possibilia"). Sometimes possibilities are also called "counterfactuals". An example would be when we make future plans, or reason about changing a past event. For example, someone might say: "had I been born a hundred years ago ...". We can make sense of such talk. We can argue whether such claims are true or false. The idea of "possible worlds", which dates back to Leibniz, is that when we say "had I ..." what we mean is that "there is a possible world where ...". The connection between this world (the "actual world") and possible worlds is truth-functional and metaphysical; it is not causal. Possible worlds have no physical impact on ours, but they do sanction attributing truth-values to counterfactual sentences. There is much debate over the ontological status of possible worlds: whether they exist independently of our imagination, if so in which sense, and how many of them there are, whether they are abstract or concrete.

Routley and Graham Priest apply Meinong's distinction between being a nonexistent object and being an existent object to mathematics. Mathematical theories are possible but not actual. They believe that the actual world exists, and that physical objects exist. They do not believe that abstract mathematical objects exist. Nevertheless, they are objects. Moreover, they are possible objects, so they inhabit a possible world. They are possible simply in virtue of having properties, and they adopt Meinong's criterion for being an object. Anything with properties is an object. Routley and Priest consider that mathematical objects are possibilia.

Meinong, Routley and Priest are conceptually generous. They want to leave room for impossible objects and even contradictory objects. Impossible objects are objects like the mountain made of gold. This is impossible in the sense of there simply not being that much gold in the actual world we inhabit. A mountain made of gold is logically possible, but not actually possible: that is, it is not possible using the actual world as a reference. A mountain made of gold has properties: namely, of being a mountain, and being made of gold. Therefore it is an object. Since it is an object it is (at least) a possible object, so there are possible worlds which have this mountain: not the actual world, but merely possible worlds.

Meinongian philosophy is yet more radical. There are objects for Meinong, and also for Routley and Priest, that are contradictory. Such an object has properties, for example, that of "being a contradictory object". An example of a contradictory object is a round square, or an object that is both wholly blue and not blue at the same time and place. We would like to say that a contradictory object is "logically impossible", but then we would have a problem with how we are expressing ourselves. It is not clear at all that we want to have impossible worlds, for they are impossible. So, by definition of "impossible" there should be no corresponding object or world.

Tidying up the language, mountains made of gold are not possible, given a reference to our actual world. We shall call these "actually impossible". They are logically possible in the sense that logic does not find any problem with attributing both the property "made of gold" and "is a mountain" to one and the same object. Logically impossible objects, or contradictory objects, such as our blue and not-blue objects are more problematic, for if the object is contradictory then it will spread its contagion of contradiction to any object it touches through *ex falso quod libet* (the lemma of classical logic that says that from a contradiction anything follows). If an object is logically impossible it should not occupy a possible world since contradictions are logical impossibilities, so there should be no corresponding possible world, even if we do say that they are nonexistent.

Routley and Priest developed a way out of this conundrum by developing paraconsistent relevant logic (Priest 2003: 14). Relevant logics are logics that insist on there being some sort of connection between the premises and conclusion to an argument. In particular, relevant logics block the classically legitimate inference from a contradiction to anything at all: *ex falso quod libet*. It was this inference, allowed in classical logic, that got Frege into such trouble with his formal system, and similarly with other formal systems with paradoxes. Having a relevant logic underpinning the "conceptual space" or "logic" of possible worlds stops the ubiquitous spread of inconsistency. To stem the spread of inconsistency we need two more considerations. One is that "existence" be turned into a predicate, as opposed to a quantifier, and as such be attributable only relative to a world. For example, Hamlet exists relative to the context of the play by Shakespeare so in a possible world where the play is "real", the physical objects in the play world are physical. In particular, Hamlet is physical. He has mass, location and duration, at least while he is "alive". Thus, Hamlet is physical in the play world of Shakespeare's play *Hamlet*. In relevant paraconsistent logic, the existential quantifier does not disappear, but it is interpreted differently from the classical existential quantifier. The relevant paraconsistent existential quantifier is read "some". So "$\exists x[Fx]$" is read "Some x has the property F". The existential quantifier does not commit us to the existence of the object being quantified over. There might be no x, even if some x has F. This is alright because it might be a nonexistent x that has the property F. For example, if F is "has the property of being a unicorn", then there is no x that has this. Nevertheless, some (nonexistent) x might. The existential quantifier only serves to quantify over objects. "Some objects are white" will be true of both existent horses and nonexistent unicorns. We need the further existential predicate, "E", not quantifier, to assert that an object exists, Ex, (in a world). Thus, a distinction is drawn between the existential quantifier and the existence predicate.

The other consideration we need explains how it is that objects get their characterization. For this we need a "characterizing principle" (CP). The principle is this:

CP An object has the properties it is characterized as having and any characteristics that follow from those properties.

This merits elaboration. "Any" means any as allowed by the relevant paraconsistent logic. Paraconsistent logics are logics that allow contradictions.

Contradictions are analysed as being both true and false. In particular, consider the liar paradox. The sentence "this sentence is false" is both true and false according to the paraconsistent logician.[20] So, there are some objects, occupying worlds, that are characterized as having a property and its opposite. Take for example our blue but not-blue object. It has the property of being blue all over at a given time and place. It has another characteristic of having no blue on it at the same time and place. These are two characteristics belonging to the object. Since it has characteristics, it is an object. Since it is an object, it occupies a world. The world is one that has impossible objects. Note that this odd object is neither actual nor does it exist. This is because, in Priest's Meinongian philosophy of mathematics, the only existent world is the actual world. It follows that anything that does not exist in this world does not exist at all, including possible worlds themselves. This does not prevent objects from occupying nonexistent possible worlds.

Returning to mathematics, a contradictory classical system is trivial, in the technical sense that anything follows from the contradiction through *ex falso quod libet*. Thus "2 + 2 = 19", "2 + 2 > 8" and "the internal angles of any triangle add up to 230 degrees" are all truths of a contradictory classical formal system of mathematics. They are also all false in a classical system. If the underlying logic of the system is not classical but, rather, relevant and paraconsistent, then *ex falso quod libet* inferences are not allowed, so there is no spread of inconsistency. Not every sentence of the theory is both true and false: only some paradoxical ones are. The limitation is often exercised through the rules of inference of the system, which ensure relevance. So, simplistically, from P and $\sim P$ we can infer P, $\sim P$ or $\sim\sim P$; we cannot infer Q. Details can be found in the Appendix.

How will this help with the philosophy of mathematics? Why should we want to accommodate contradictions? There are pairs of mathematical systems that contradict each other. There are also mathematical systems that are internally contradictory. They tend to be unsuccessful, but they are nevertheless part of mathematics. We want to discuss them. The objects of those mathematical theories are objects in contradictory worlds: worlds with pairs of sentences that are both true and contradict each other; P and $\sim P$ are both true (and false). We shall leave these aside for now and just concentrate on pairs of consistent systems. An example of a pair of mathematical systems that are mutually contradictory are Euclidean geometry and projective geometry. The two geometries differ over the truth of Euclid's fifth postulate, so there are some theorems of Euclidean geometry that are denied by projective geometry. In particular, in projective geometry parallel lines do meet (at infinity); in Euclidean geometry parallel lines never meet.

The Meinongian philosophy of mathematics will tell us that there is a possible world where Euclidean geometry is true and a possible world where projective geometry is true. There are possible worlds corresponding to each mathematical theory. Moreover, the Meinongian allows us reasonably to contemplate putting the two theories together, and discovering that that possible theory (or possible world) is inconsistent. Thus, Meinongian philosophy of mathematics is more liberal than the platonist who insists that one of the two theories of geometry is correct. The Meinongian philosophy of mathematics is more liberal than Hellman's modal structuralism because the underlying logic is paraconsistent, not classical.

Let us specify what "each" means above, where we talk about "each mathematical theory". It will certainly include all the mathematical theories we study today. "Each" will also include past theories we have stopped studying, and possible theories: ones we have not yet started to study. "Each" could (depending on how much stomach one has for paraconsistency) also include inconsistent systems of mathematics. In so far as one thinks that contradiction is the same as logical impossibility, there will be impossible worlds that have inconsistent mathematical systems. There will also be inconsistent possible worlds; recall that these do not exist. Nevertheless they have characteristics, so we can talk about them. Meinong's nonexistent objects are transformed into Routley's and Priest's nonexistent possible worlds, some of which are mathematical theories. A theory might be consistent or inconsistent. One inconsistent world is Frege's world of second-order logic; a contradiction was derivable in the formal system. The formal system is classical, so allows *ex falso quod libet* inferences. The formal system has a number of characteristics. Therefore there is a corresponding mathematical world. The virtue of this way of organizing the system of possible worlds is that it allows us to make sense of our talk of inconsistent formal systems, such as Frege's. We do study and discuss Frege's formal system, and we are not talking nonsense when we do so. When we study Frege's formal system, we usually try to salvage parts of it, or modify it in some way. Nevertheless, we are discussing "it", and "it" is inconsistent and trivial.

How do we stop the spread of inconsistency from one world to another (consistent) one? Through relevant paraconsistent logic. This is used to govern the "universe of possible worlds". That is, the very organization under which all these worlds find themselves is a relevant, paraconsistent organization. If a world is inconsistent, it does not follow that every other world is inconsistent. Frege's inconsistent logic is hermetically sealed from others.

The advantage of this view is its generous pluralism. The Meinongian philosopher of mathematics will not censure any mathematical theory, however "crazy" it is. This does not prevent the Meinongian from saying that one theory is not as useful as another theory, or that one theory is inconsistent with another theory, or that more mathematicians like or believe in one theory over another. However, all theories have their place in a possible world.

The objection to this is that we do not want our philosophy of mathematics to be so liberal. Normally, what we understand by a philosophy of mathematics is a theory of what counts as *successful* mathematics, not just any combination of mathematics-like elements. The philosopher of mathematics does not have to account for, or show an interest in, bad (i.e. inconsistent) mathematics. Similarly, if we give a philosophy of science we do not give a philosophy of anything that can assume the garb of science. Instead, we want to say that only some of our enquiry is good and successful, and this is what counts as science. Philosophers give a theory of the good practice, not of the good and the bad practice all mixed together. The divide between this philosophy and the others we have examined in this book is quite deep, for it involves an explanation as to what counts in giving a philosophy of some topic. We saw a little of this probing enquiry in the section on Husserl. Meinongian philosophy of mathematics is another way of radically departing from other positions in the philosophy of mathematics.

9. Lakatos

Imre Lakatos's philosophy of mathematics is difficult to situate among the standard philosophies of mathematics. He is interested in the process of mathematical discovery. How do we learn more mathematical theories, and what are the more instructive lessons? We might say that the way in which we learn more mathematics is that we prove more and more things. Proofs teach us more mathematics, and they take us from things that we knew to new things that we did not previously know. Interestingly, Lakatos only partly agrees with this. He thinks that we do learn a little from proofs, but we actually learn a lot more from dis-proofs. If we think about it, this makes sense. If we get a proof of some theorem we have been trying to prove for a while, then this just confirms our thinking. Everything falls into place where we expected it. If we find a dis-proof, then this causes us radically to rethink our position. Our instincts are shown to be wrong; the theory is not so predictable. This makes for a deeper change in our mathematical knowledge than a straightforward, confirming proof.

Lakatos railed against what he called "static rationality" (Larvor 1998: 19). This is the view, held by any realist of mathematics, that mathematics is a fixed body of knowledge, got at by means of fixed rules of inference: a static logic. This is not to say that Lakatos disliked logic; his objection was against the philosophical import attributed to logic, by logicists in particular. Logicists believe that their logic, be it second-order logic or type theory, is our only means of really justifying the body of mathematical truths, and that therefore mathematics (or whatever part can be justified by logic) is essentially logic.

This imports the static reasoning of logic to mathematics. Moreover, Lakatos was not convinced that there was such a fixed, or static, body of mathematical truths.

In order for there to be such a body of truths, mathematical language would have to be fixed. The typical realist's view is that "triangle" means triangle and it has done so from the first use of the word to today. Instead of this view, Lakatos believes that mathematical language develops. As we learn more about geometry, our understanding of "triangle" changes, and the meaning of the word changes. A perfectly adequate Euclidean characterization of a triangle as a three-sided closed figure is naive, and adequate only until non-Euclidean geometries were developed. "Triangle" is, today, ambiguous. Do we refer to triangles on the Euclidean plane, triangles on the outer surface of a sphere or triangles on the inner surface of a sphere? Are these all triangles? From how we use the term, it seems that "triangle" not only refers to a number of different types of triangle, but also different types in different contexts (on different sorts of surface, which change the properties of the triangle). These changes in use of the term "triangle" changed when the mathematical community learned about non-Euclidean geometries, and decisions had to be taken over what was essential to being a triangle. Is a triangle essentially a three-sided figure? Is a triangle essentially a three-sided figure with the sum of the interior angles adding to 180 degrees? Do the sides of the figure have to be straight, and what do we mean by "straight"? A triangle on the outside surface of a sphere has a sum of interior angles adding to more than 180 degrees (Fig. 14). (The lines connecting the angles are the shortest possible, and curve along the surface of the sphere.) These questions had to be resolved by appeal to our Euclidean notion of "triangle", and what prompted all these questions was the refutation that Euclid's fifth postulate was derivable from the other postulates. When it was discovered that the fifth postulate was independent, this profoundly overturned out thinking about geometry. We learn much more from this sort of "refutation" proof than from a straight proof of a theorem of geometry.

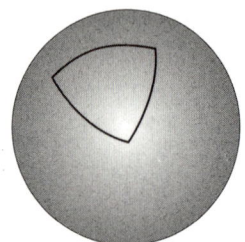

Figure 14

Lakatos observed that when mathematicians, raised on a steady diet of Euclidean geometry, are faced with non-Euclidean geometry, they have to come to some agreement as to what terms like "triangle" mean (essentially).[21] It took a while for the vocabulary to gel. The process of refining our understanding consists in confronting alleged counter-examples to definitions, equations, rules or axioms. Lakatos further observed that there are different ways of reacting to a counter-example:[22] modify the definition, say, to include the new example; reject the definition altogether as hopeless; add an "exception" clause to the original definition; or make up some new words to deal with the separate cases, so we might, for example, have to distinguish "triangle-on-the-Euclidean-plane" from "triangle-on-the-surface-of-a-sphere". Reading older textbooks in mathematics is highly instructive on which terms have or have not yet gelled. We see some terms being defended, and others taken for granted.

Lakatos found the static style of writing adopted by mathematicians to be highly deceptive.[23] Not only does this style remove any motivation for the reader to learn the piece of mathematics, but it also gives a false impression as to what mathematics really is (Larvor 1998: 27). Mathematics, for Lakatos, is an evolving and changing body of knowledge. The change occurs in the direction of greater sophistication and refinement.

But now we have a problem. We have to determine whether there is any distinction to be drawn between mindless developments in false refinement of words and concepts, and progress in true, or good, understanding, for it is possible for someone to introduce a distinction that leads us astray. Profound refutation-type proofs, or proofs of limitations of a formal system, certainly generate a great deal of activity, but is all the activity legitimate, and how do we judge whether it is? If there is a distinction to be drawn, then we have to give some indication as to how it is that we can favour one type of activity over the other. From at least some of the things Lakatos writes, it is clear that he, at least at times, does want to draw such a distinction. If he does not, then Lakatos is simply describing a psychological, subjective, descriptive view of mathematics. Mathematical development is not then viewed as a rational process but simply as a human activity comparable to making up games.[24] Usually, we do not make up games to come closer to generating the ultimate game, for there are many games on a par with each other. Nor do we strive to find out truths about games, or to discover truths that we cannot know about by other means. Instead, we make up games either for amusement, or as a response to a psychological need. We do not do science when we invent games. Any notion of progress towards a goal, outside the game, is misguided. There is no ultimate body of games. There just is a body of games, to which we add to as we see fit. This begins to sound like formalism. Lakatos "hopes to examine the growth of mathematics *philosophically.* That is, he hopes to depict it as a rational objective process" (*ibid.*: 21). Subsumed under this declaration there has to be a sense of progress. Thus, there is a sense of correct and incorrect direction. "For Lakatos, progress in mathematics means that the concepts employed by mathematicians approximate more and more closely the objective structure of mathematical reality" (*ibid.*: 25). The deep problem consists in recognizing the "objective structure" when we see it. We have to know which are good directions and which are bad directions. "Even if we had a guarantee that mathematics will always progress (which we do not), there would still be no knowing which direction this progress would take" (*ibid.*: 29).

Maybe mathematics does not progress towards an "objective" structure, but rather progresses only in the sense of ever greater refinement in our mathematical concepts and understanding. If this is the case, then there is no mathematical justification for preferring one development from another provided it is "more refined" under some understanding of "more refined". There is no rationale behind favouring one result over another, of finding one more important than another; and yet, it is this, for Lakatos, that is the source of motivation, discovery and mystery in mathematics. We strip mathematics of significance if it is just an idle activity of theorem or refutation production. We might as well just set the computers whirring to produce ever more theorems. Of course, this would not do, since mathematics is for human beings, not computers; and the computers do not have the creativity to change a definition once it is given to them.

Lakatos did have a certain amount of faith in the "rational process". Lakatos refused to give a logic to this process. The problem is that, in so far as he resists formalization of that reasoning process, he forfeits any decisive ability to know whether we are engaged in properly directed reasoning or some deceptive imitation of reasoning that will, ultimately, lead us astray. Maybe Lakatos really should go all the way and say that mathematics is just a human activity. We engage in it. The only discernment of good from bad mathematics is determined by mathematicians and, in the end, the decisions come from a mixture of factors, some of which are social. There is no mathematically objective body of knowledge. Ultimately, there is no mathematical, or logical, justification for the activity, or account for the directions it takes. Ultimate justification rests in the human institutions of mathematical activity: journals, textbooks, mathematics departments, conferences. Engaging in the activity is what is important, rather than the content or focus of the activity. The focus, or content, only delineates mathematical activity from others. Partly inspired by Lakatos, David Corfield has a much more pluralistic and piecemeal approach to the philosophy of mathematics, thinking that it is unrealistic to try to offer a philosophical position to account for, or justify, all of mathematics. Instead, the philosopher should concentrate on local problems within mathematics. Maybe this is a viable new direction in the philosophy of mathematics.

Appendix
Proof: *ex falso quod libet*

This is a classical proof that from a contradiction, anything follows: $p, \sim p \therefore q$. It can be set out as follows:

{1}	1.	p	Premise
{2}	2.	$\sim p$	Premise
{3}	3.	$\sim q$	Assumption for a *reductio ad absurdum* argument
{1, 3}	4.	$p \,\&\, \sim q$	1, 3 &-introduction
{1, 3}	5.	p	4 &-elimination
{1, 2, 3}	6.	$p \,\&\, \sim p$	2, 5 &-introduction
{1, 2}	7.	$\sim\sim q$	3, 6 *Reductio ad absurdum*, discharging 3
{1, 2}	8.	q	7 Double negation elimination

This proof is not valid intuitionistically or paraconsistently, for *reductio ad absurdum* and double negation elimination are not rules of inference in either system. We can avoid using *reductio ad absurdum* by either using the rule of conditional proof together with *modus tollens*, or by using disjunctive syllogism.

Here is the proof using disjunctive syllogism, which is intuitionistically valid:

{1}	1.	p	Premise
{2}	2.	$\sim p$	Premise
{1}	3.	$p \vee q$	1 ∨-Introduction (weakening)
{1, 2}	4.	q	2, 3 Disjunctive syllogism

The intuitionist allows disjunctive syllogism. The paraconsistent logician will not allow disjunctive syllogism as a rule of inference (Priest 2001: 151). For the paraconsistent logician, it is possible to have an inconsistent sentence, such as the liar sentence, as an axiom or derived theorem, without the whole theory becoming trivial. That is, it will not be the case that every well-formed formula of a theory containing a contradiction will be derivable. The theory is then inconsistent (contains a contradiction) but non-trivial (not anything is derivable). This is because paraconsistent logic allows truth-value gaps and truth-value gluts. A truth-value gap occurs when a well-formed formula of a language gets no truth-value assigned to it. A truth-value glut occurs when both "true" and "false" are assigned to the same sentence. An example of a sentence that will enjoy a truth-value glut is a paradoxical sentence such as "This sentence is false". To show the invalidity of the inference $p, \sim p \therefore q$, take p to be both true and false. Take q to be false. Then both premises are true, but the conclusion is false (*ibid.*). Both premises are also false, but the definition of validity is worded exactly the same as in classical logic: if the premises are true, so is the conclusion. Because paraconsistent logic blocks *ex falso quod libet* inferences, it is possible to have theories that contain inconsistencies but that are not trivial.

Glossary

***a posteriori* truth** A proposition is an *a posteriori* truth if it can only be known by appeal to sense data. These are empirical truths. *A posteriori* truths are contrasted to *a priori* truths.

***a priori* truth** A proposition is an *a priori* truth if and only if it is not necessary to have any particular sense experience in order to recognize that it is true. Truths of logic and many (arguably all) analytic truths are considered to be *a priori*.

abstraction principle In the literature, this is also called "contextual definition". Examples are: basic law V; the parities principle; Hume's principle (the numbers principle). The necessary features of an abstraction principle are that it should be a second-order universal formula. Inside the scope of the second-order universal quantifiers we find a biconditional expression. On one side of the biconditional there is an identity, on the other side there is an equivalence relation. The numbers principle is: $\forall F \forall G(Nx{:}Fx = Nx{:}Gx \leftrightarrow F \approx G)$. This is read: for all concepts F and for all concepts G, the number of Fs is identical to the number of Gs if and only if F and G can be placed into one-to-one correspondence. One-to-one correspondence is an equivalence relation.

acquisition argument The acquisition argument against the decoupling of truth from understanding or knowledge is used by anti-realists against realists. The realist claim is that there are verification-transcendent truths, that is, truths that lie beyond our abilities to verify or experience them. The anti-realist asks: how it is possible for there to be truths that we cannot verify or experience, since it then seems that we cannot acquire understanding of that truth? The purported (candidate verification-transcendent) truth is then devoid of content, and therefore meaningless.

actual world The actual world is the real world we live in. This is contrasted to merely possible worlds.

analytic truth A truth is analytic if and only if it is true in virtue of the meaning of the sentence. A celebrated example of an analytic truth is: "All bachelors are unmarried men". This is not a statistical coincidence; it is true in virtue of meaning. If a proposition is an analytical truth it cannot be a synthetic truth.

ante rem "*Ante rem*" is translated as "before reality". The term is most famously used when discussing properties. Properties, such as colours, are *ante rem* just in case they exist independent of whether or not there are any real or actual objects that happen to have the property. Applied to structures, the idea is that a structure exists independently of whether or not there are any mathematical objects that happen to satisfy the structure. *Ante rem* is contrasted to *in re*.

anti-realist Unless he is a "global anti-realist", an anti-realist is an anti-realist about a particular discourse or subject matter (and a realist about others). If one is an anti-realist about numbers, then one thinks that numbers are created by us: their existence depends on our having created them, or on our having the ability to create them (under some sense of "able"). Often this is expressed as: the truths about numbers are "epistemically constrained". That is, there are no verification-transcendent truths about numbers.

axiom An axiom is a basic truth of a system (or theory). From axioms, together with some rules of inference, we can derive theorems. Examples of axiomatic systems in mathematics include Euclidean geometry and Zermelo–Fraenkel set theory. Not all of mathematics is developed axiomatically. However, it is thought that if we can find the axioms of a theory, we can gain a more precise understanding of the theory. We are also less likely to make mistakes in our proofs of purported theorems of the theory.

axiom of choice There are many versions of the axiom of choice. For the purposes of this book it is enough to have a rough idea. The axiom of choice guarantees that there is a "choice function" for any set A composed of other sets. The choice function will pick out one member from each set (which is a member of A). The truth of the axiom of choice is independent of Zermelo–Fraenkel set theory. For this reason, there is another set theory called "Zermelo–Fraenkel set theory with choice". The truth of the axiom of choice is disputed by constructivists.

axiom of infinity An axiom of infinity is one that guarantees the existence of an infinite set. A formal system might have several axioms of infinity to guarantee the existence of different sorts, or sizes, of infinity.

basic law The term "basic law" was coined by Frege, and refers to a law, or axiom, of logic. This is more basic than an axiom of a particular mathematical theory.

bivalence A logic is bivalent if, in its semantics, it has two truth-values: true and false.

cardinality Cardinality is the measure of the size of a set, answering the question: how many members does it have?

causal In philosophical literature, "causal" tends to be restricted to physical causation, as opposed the broader explanatory causation (whatever can follow the word "because" in English).

class A class is a gathering of objects under some concept or predicative term. All sets are classes, but not all classes are sets. Those that are not sets are proper classes. Proper classes are not obtained from the axioms of set theory. They are obtained by thinking up a concept that we cannot construct from the set-theoretic axioms. Intuitively, proper classes are "too big" to be sets. Examples are: the set-theoretic hierarchy, all the ordinals and all the cardinals.

compactness The compactness theorem is true of some logics (or mathematical theories). If a theory is compact then a formula A is valid (always true) in the theory if and only if A is valid in some finitely axiomatized part of the theory. Second-order logic is not compact because there are some valid statements that cannot be proved using a finite number of axioms.

compositionality "Compositionality" refers to the structure of languages. A language is compositional if and only if the meaning of larger units is completely analysable in terms of the meaning of smaller units. More mundanely, the meaning of sentences can be completely understood in terms of the meaning of the words in the sentence together with how those words are ordered (so we can refer to the grammar of the language). A language would not be compositional if and only if the meaning of large units, such as sentences, went beyond the meaning of the parts.

concrete object A concrete object is a physical object. It has location and mass.

consistency A theory is consistent if and only if it is not possible to derive a contradiction in the theory.

constructive logic There are many constructive logics. What they have in common are that they are motivated by some sense of epistemically constraining truth through the logic. To meet those ends they reject some of the principles of classical logic. They will reject some combination of: the rule of double negation elimination; *reductio ad absurdum*; certain versions of *modus tollens*; the axiom of choice; existential elimination; the law of bivalence; and the law of excluded middle. Intuitionist logic is a type of constructive logic.

context principle See "abstraction principle".

continuum problem The continuum problem is deciding whether or not the following "continuum hypothesis" is true (in set theory): $2^{\aleph_0} = \aleph_1$. That is, the problem is to decide whether or not $\aleph_1$ (which is the next size up after $\aleph_0$) is got at by raising 2 to the power of $\aleph_0$. The problem was posed by Cantor. It was resolved much later, when mathematicians discovered that the hypothesis is independent of Zermelo–Fraenkel set theory with choice.

contradictory object A contradictory object is one that has properties that preclude each other. An example is a round square or a hat that is only blue all over at the same time as being only red all over. Contradictory objects do not exist, and yet we can reason about them. As a result, Meinong proposes that they are treated as nonexistent objects.

counterfactual A counterfactual is a fact that does not actually obtain. These are facts that could have been, or that depend on some prior conditions that do not happen to obtain.

decidable A logic, or mathematical theory, is decidable if and only if our (proof) test for theorems will always give us a definite answer in a finite number of steps. Decidability concerns the efficacy of our proof system. A theory is undecidable if and only if there are some truths of the theory that cannot be proved to be true by the proof system in a finite number of steps (some proofs will continue forever).

density Density refers to series of numbers. A series of numbers is everywhere dense if and only if between any two there is a third. The rational numbers and real numbers are everywhere dense. The natural numbers, and integers are not.

disjunctive syllogism The logical rule of disjunctive syllogism is that if you have a disjunction and the negation of one of the disjuncts, then you can infer the other disjunct (is true).

empty set The empty set is the set with no members. It is a subset of every set (by the definition of "subset"), but is not a member of every set.

epistemology Epistemology is the study of how we know, or what our knowledge consists in, concerning some area of study.

equivalence Equivalence is a looser relation than identity (or equality). Two objects are equivalent if and only if they are the same in some respects. If they are the same in all respects then they are "the same" object, that is, "they" are identical: we have only one object.

extension of a concept The extension of a concept is all the objects that fall under the concept. An extensional definition is one that lists the objects. In contrast, an intensional definition is one that gives us a means of picking out the objects. Most mathematicians think that all of mathematics is extensional. That is, which expressions we use to pick out objects is irrelevant to mathematics. Provided we pick out the same objects, no mathematician cares about the particular expression. For example, the expression "2 + 2" has the same extension as the expression "8 − 4".

free logic The "free" in free logic refers to "free of ontological commitments". A free logic is designed to accommodate reasoning over nonexistent objects, for example, fictional objects. There are several free logics. Typically, they take issue with the classical rule of existential instantiation, or existential elimination: that from an existentially quantified sentence we can directly infer a named version of that sentence (provided the name is free – not used already in the proof), where the name refers to an object in the domain. $\exists x(Fx)$ therefore Fa, where "a" is a name for an object in the domain, and is therefore, guaranteed to exist.

gapless We use the term "gapless" in two different ways in the text. One is when we refer to the continuum as a gapless line. That is, the continuum is a smooth line, with no points, or numbers representing those points, missing. Frege uses the term "gapless" to refer to proofs. A proof is gapless just in case the reasoning is completely tight. Every step in the proof is accounted for either by an axiom or by a rule of inference from previous lines in the proof. The virtue of a gapless proof is that we can be sure that no unexamined presupposition has crept into the proof.

higher-order logic A logic is a "higher-order" logic if and only if it allows quantification over higher-order variables. That is, it will allow quantification over predicates, relations and functions. This is contrasted to first-order logic, which only allows quantification over objects.

impredicative definitions A (non-impredicative) definition should pick out an object, and it should express itself in terms different from the term being defined. In contrast, an impredicative definition is one that uses the terms being defined in order to give the definition. In some way the definition is then circular.

in re "*In re*" is translated as "in reality". The term is most famously used when discussing properties. Properties, such as colours, are *in re* just in case their existence depends on there being some objects that have the property. Applied to structures, the idea is that a structure exists only if there are mathematical objects that satisfy the structure.

indispensability arguments Indispensability arguments are used by naturalists to defend some parts of mathematics. The idea is that the part of mathematics that is indispensable to physics, or our other scientific theories, is good, vindicated mathematics. The rest of mathematics is suspect.

integers Integers are the whole numbers and all the negative whole numbers: ...−3, −2, −1, 0, 1, 2, 3...

irrational numbers Irrational numbers are numbers that cannot be expressed as a fraction with whole numbers as numerator and denominator. They have to be expressed as numbers with a decimal. The numbers after the decimal never repeat in the same finite pattern. The "same finite pattern" is called a "period": it can have any finite length. An example is: 345. A number that has the period 345 will continue infinitely after the decimal point with ... 345345345345.... Any number with a period can be expressed as a fraction, so is rational. The number π is a famous irrational number.

law of excluded middle The law of excluded middle is a syntactic law, or axiom. It states that for any well-formed formula either it or its negation holds. The law of excluded middle is often rejected by constructivists. They add that a well-formed formula also has to be constructed (by the constructive rules) in order to hold. The semantic counterpart of the law of excluded middle is bivalence. Intuitionists accept bivalence but reject the law of excluded middle.

limit ordinal A limit ordinal is an infinite ordinal with no immediate predecessor. ω is the first limit ordinal. It follows all the finite ordinals, so is infinite, and it has no immediate predecessor.

logical object A logical object is one that logic says exists. If logicism is right, then numbers are logical objects.

Löwenheim–Skolem property The Löwenheim–Skolem property pertains to mathematical theories or logics. The theorem is: if T is a countable theory having a model then it will have a countable model. A theory is countable if and only if its language is of size $\aleph_0$. For example, the language of propositional logic is countable, since it has five logical connectives and an infinite number of proposition letters. If the Löwenheim–Skolem theorem is true of a mathematical theory or logic, then that theory, or logic, has the Löwenheim–Skolem property.

manifestation argument The manifestation argument is deployed by anti-realists against realists, and concerns our understanding of a truth. The anti-realist believes that it makes no sense to attribute understanding of a truth to another person if that person cannot manifest her understanding of that truth. The anti-realist asks how it is possible for that person to have understanding. Note that "manifesting" does not necessarily require a full explanation, but it does require correct(able) use. If a person claims to have understanding of a truth concerning a concept, but cannot give some indication of how to use those words (even by example) then we cannot attribute understanding to that person. The purported truth is verification-transcendent, and therefore meaningless.

modus ponens *Modus ponens* is a rule of inference. It says that if we have a conditional statement, and the antecedent of that conditional, then we may infer the consequent of the conditional.

modus tollens The classical version of *modus tollens* is that if we have a conditional and the negation of the consequent, then we may infer the opposite of the antecedent. The constructivist is more careful. He says that if we have a conditional and the negation of the consequent we may infer the *negation* of the antecedent. If this happens to be a doubly negated formula, then we need to use double-negation elimination to get rid of the double negation, and the rule of double-negation elimination is rarely endorsed by the constructivist.

natural deduction Natural deduction is a formal syntactic system of deduction, where we reason from premises using rules of inference to the conclusion. Natural deduction is distinguished from tree proofs or table proofs, which are semantic.

non-classical logic A logic is non-classical if it rejects some of the principles of classical logic. See "constructive logic".

non-standard models of arithmetic Non-standard models of arithmetic are only allowed in first-order arithmetic. If we move to second-order arithmetic (where the axiom of induction includes quantification over properties), then there are no non-standard models. A non-standard model is a series of numbers that at the beginning will look exactly like the natural numbers but, "after" all the finite numbers, the number line looks different than what we see with the standard ordinals. There are many non-standard models. An example will have all of the finite ordinals, and then the infinite ordinals as per our regular theory, but there will be an infinite number of copies of infinite ordinals.

normativity "Normative" is contrasted to "descriptive" and sometimes "prescriptive". A theory, or principle, is normative if it sets a norm. This is not a statistical norm; rather, it is a standard. A descriptive theory, or principle, simply describes what is. A prescriptive theory, or principle, prescribes what we ought to do.

noumenal world "Noumenal world" comes from Kant. The noumenal world is the world as it is in the raw: the world as it really is. We do not directly interact with it. Instead, our world is the phenomenal world. The phenomenal world lies between the noumenal world and our concepts and experiences. We have contact with the phenomenal world, which is the noumenal world shaped by our concepts and our abilities to sense the world.

one-to-one correspondence More accurately, this should be written "one-to-one and onto". Two sets can be placed into one-to-one (and onto) correspondence if and only if every member of one set can be matched with exactly one member of the other set, and *vice versa*. We say that two sets that can be placed into one-to-one (and onto) correspondence are of the same size.

ordering relation An ordering relation imposes an order on a set of objects. Examples of an ordering relation are: "is greater than"; "is taller than"; and "is older than".

ordinal An ordinal is a number used to give a place in an order. For example, "first", "seventh" and "eighteenth" are all ordinals.

phenomenal world See "noumenal world".

possible world Possible worlds are places where possibilities are played out. The term "world" in this case is not restricted to a planet the size of earth. Rather, these will often include whole universes. Whenever we use the locution "it is possible that ..." the idea is that what makes the locution true is that there is a possible world where "that ..." happens. In the literature, there are many questions concerning the ontological status of possible worlds: whether they really exist, are abstract, whether past ones exist and so on. They are not causally linked to our world but, by supposing them, we endorse counterfactual statements: we make them truth-apt.

powerset The powerset of a set is the collection of all the subsets of a set.

prescriptive See "normative".

prime pairs Prime pairs are pairs of prime numbers that are separated by only one even number. Examples include <3, 5>, <5, 7>, and <11, 13>.

proposition A proposition is a fact referred to by means of a declarative sentence. A proposition is truth-apt.

propositional logic Propositional logic is sometimes called "sentential" logic. The logic is simpler than first-order logic or higher-order logics. Propositions are taken as an irreducible unit. So any declarative sentence is replaced by a proposition letter. The logical connectives are also part of the vocabulary. First- and higher-order logics are more sophisticated, since they allow us to analyse propositions, and not take them as irreducible.

quantifier-free elementary arithmetic "Quantifier-free" means "without quantifiers", so quantifier-free elementary arithmetic is the arithmetic concerning particular numbers. $2 + 5 = 7$ is a true formula (theorem) of quantifier-free elementary arithmetic. The formula expressing the commutativity of addition $\forall x \forall y (x + y = y + x)$ is not a formula of quantifier-free elementary arithmetic. "Elementary" refers to it's being first-order.

reductio ad absurdum *Reductio ad absurdum* is a rule of inference of classical logic. It states that if a contradiction follows from a formula then that formula is incorrect. The rule runs: assume (for the sake of argument) that P, where P is some formula. Prove a contradiction from P. You may then infer that it is not the case that P.

relevant logic A relevant logic is one that will block *ex falso quod libet* arguments. That is they reject the classically valid arguments with contradictory premises. It is classically valid that from a contradiction anything follows.

second-order logic Second-order logic allows quantification over second-order variables. There are many second-order logics.

singular term "Singular term" is a grammatical expression. A singular term refers to only one object. In contrast, a general term might refer to several objects. "Is an object in the room" is a general term (provided there are several objects in the room). "Is a moon of Earth" is a singular term.

size of set The size of a set is its cardinality.

soundness of a logic A logic is sound if and only if every syntactically proved argument is truth-functionally valid. That is, the syntactic proof system will not allow us to generate a false conclusion from true premises.

spatiotemporal intuition Spatiotemporal intuition is postulated by Kant to explain how it is that it is possible for us to have experiences of the external world at all. We do so, by participating in, or by making use of, spatiotemporal intuition. This allows us to organize objects in the world spatially and temporally. More interesting is Kant's idea that it takes spatiotemporal intuition to enable us to understand geometry and arithmetic. This makes the truths of geometry and arithmetic synthetic, for Kant.

supervenience Roughly, a property supervenes on an object if and only if the property would change if the object were to change. There are strong and weak formulations of supervenience. For example, beauty supervenes on a painting because if we were to alter the painting, then the beauty would disappear (or would be a different example of beauty).

synthetic truth A synthetic truth is a proposition that is true by bringing independent concepts together. For Kant a synthetic truth might be *a posteriori* or it might be true because it appeals to spatiotemporal intuition.

tautology A tautology is a sentence that is always true. Logical truths are tautologies. Any sentence of the form p implies p, where p is a proposition, is a tautology.

temporal logic A temporal logic is one that includes temporal operators. Temporal logics are designed to set norms for reasoning over time.

truth-apt A sentence is truth-apt if it is a candidate for getting a truth-value. Questions are not truth-apt; nor are nonsense sentences.

universal quantifier The universal quantifier belongs to first-order logic and any logic higher than first-order logic. The symbol for the universal quantifier is $\forall$. It is used to represent the locution "for all" or "all" in English.

validity An argument is valid if and only if, given true premises, the conclusion is also true. Another way of putting this is that an argument is logically valid if and only if it is impossible for the premises to be true and the conclusion false. A sentence, or formula is valid just in case it is always true, that is, if it is a tautology.

verification-transcendent truth A truth is verification-transcendent if and only if it lies beyond our abilities to verify it. What counts as "our abilities to verify" is up for debate and, in accordance with this, philosophers will be inclined towards considering different propositions as examples of verification-transcendent.

Notes

Chapter 1: Infinity

1. See W. Sieg, "Mechanical Procedures and Mathematical Experience", in *Mathematics and Mind*, A. George (ed.), 71–117 (Oxford: Oxford University Press, 1994).
2. For example see Aristotle's discussion of infinity; Aristotle, *Physics*, Book 6, Chapter 9, 239b–240a.
3. The *Gilgamesh Epic*, a series of legends and poems about the mythological king Gilgamesh engraved on eleven stone tablets around 1300 BCE, had its origins much earlier. It relates the story of Gilgamesh's search for eternal life.
4. *The Epic of Gilgamesh* 3.2.34–36, S. Shabandar (trans.) (Reading: Garnet Publishing, 1994), 34.
5. For information about this see the excellent article by Gregory Vlastos, "Zeno of Elea", in *The Encyclopedia of Philosophy*, vol. 8, Paul Edwards (ed.), 369–79 (New York: Macmillan, 1967).
6. In mathematics, infinitesimals are left as not further defined. As we further subdivide a space, or line, we approach a limit. However, mathematically, the process of subdiving is infinite. The very smallest divisions are the "infinitesimals". That is, we cannot, for example, divide an infinitesimal into two parts. In fact, there are a lot of conceptual problems with the calculus. See Marcus Giaquinto, *The Search for Certainty* (Oxford: Clarendon Press, 2002), 4.
7. This is referred to by Aristotle in *Physics* 239b15–18. Aristotle does not mention the tortoise explicitly. However, it was conventional to call a slow runner a tortoise. The paradox is called "Achilles and the tortoise" in other contemporary texts. We do not know what Zeno called it.
8. This is often put in terms of space and time being continuous as opposed to discrete, but this is not quite accurate with respect to the puzzle posed by the paradoxes. The rational numbers (fractions) are "everywhere dense". That is, between any two there is a third. If space and time are structured like the rational numbers then they are infinitely divisible, and this is enough to generate the paradox. The rational numbers are not enough to make a gapless line, often called the "continuum", which is numerically represented by the real numbers. The real numbers include both the rational numbers and the irrational numbers. It is a metaphysical question whether space and time are continuous, and this question is not, strictly speaking, raised by the paradoxes. In contrast, we have a conception of space and time as having smallest units, so not being

infinitely divisible. This would be represented numerically by the natural numbers or the integers. Under this conception, space and time have smallest bits of space or moments. If we work out, or decide, that space and time are not infinitely divisible then this precludes their being continuous.

9. The shift in language is made in order to reflect the shift in the text from talking of the ancient Greek debate (in which case the language of "supporter of the notion of the potential infinite" was appropriate) to the more modern debate between realists and constructivists.

10. It is unnecessary to know precisely what the term "well-ordered" means here. To satisfy the curious, the precise definition is that a series is well-ordered just in case it is ordered (by a relation such as "strictly less than", symbolized by "<"), and each subset has a least element. So, for example, the set of integers is not well-ordered, since the subset of the set of integers made up of the negative numbers has no least member. There has been much mathematical investigation into order-types for those interested in further reading.

11. The "measure" does not have to be physical; it can be mathematical. We cannot calibrate an instrument sufficiently finely to detect the difference between two very close irrational measures of some physical object, distance or time. In fact we cannot say that we have measured a particular distance between two points and found that it is of length "x", where x is an irrational number. Ponder this. However, we can order irrational numbers by the "<" relation. So, the notion of "measure" being used here is not restricted to the notion of physical measure.

12. For convenience, we are distinguishing whole numbers from natural numbers. The whole numbers begin with 0 rather than 1. We choose to do this because of the more immediate match up between the label "first" and the number 1, the first number in the series of natural numbers.

13. The observation that there are as many natural numbers as there are even numbers was first made by Galileo Galilei. To be precise, Galileo asked about numbers and their squares, and reasoned that since every number has a square, there must be as many squares as there are numbers. Cantor simplifies the example to that of numbers and their doubles. See Galileo Galilei, *Dialogues Concerning Two New Sciences*, H. Crew and A. Salvio (trans.) (Evanston, IL: Northwestern University Press, 1939).

14. Richard Dedekind, *Essays on the Theory of Numbers* (New York: Dover, 1963), 63–70.

15. The number on top of a fraction is called the numerator and the number below is called the denominator. Fractions with 0 as the denominator are called "undefined" (essentially they are infinite, under a loose sense of "infinite"). Fractions with 0 as a numerator are just 0 itself.

16. Sometimes called the Euler Number or Napier's Constant, e is an irrational constant used in working out logarithms. The number π is indispensable to geometry and is found in measurements concerning the cirle. For example, the circumference of a circle is $2\pi r$, where r is the radius of the cirle.

17. For this reason it is no wonder that this is one of the favoured proof techniques of Raymond Smullyan, who is not only an important mathematician, but also worked for a long time as a magician.

18. There are different presentations of this. I have chosen a very abstract method, but some readers might find it more helpful to envisage a list of random numbers, where some are irrational and some rational. It might also be helpful to just try to generate a list that will eventually scoop up all the numbers. The proof is a way of convincing us that not only is this a difficult task, but it is truly impossible.

19. See K. Gödel, "What is Cantor's Continuum Problem?" [1964], in *Philosophy of Mathematics: Selected Readings*, 2nd ed., P. Benacerraf and H. Putnam (eds), 483–4 (Cambridge: Cambridge University Press, 1983) and discussions in Michael Hallett, *Cantorian Set Theory and Limitation of Size* (Oxford: Clarendon Press, 1984) 2, 5, 6.

Chapter 2. Mathematical Platonism and realism

1. Socrates features as a character in many of Plato's dialogues. Socrates was Plato's teacher, and Plato recorded the dialogues between his teacher and other philosophers. It is thought that the early dialogues are loyal recordings, but that Plato's later work, such as the *Republic*, was much less influenced by Socrates, and better represents Plato's own thinking.
2. W. S. Anglin and J. Lambeck, *The Heritage of Thales* (New York: Springer, 1995), 90.
3. For a good discussion of this see Anglin & Lambeck, *The Heritage of Thales*, 89–92 on non-Euclidean geometries.
4. It might be a useful conceptual exercise to compare this to different people's experiences of a particular colour, or of pain.
5. It is not important that we cannot always rank any pair of drawn triangles. It suffices that there are pairs of triangles that we can rank. The difference is subtle.
6. This is a little ironic since Plato chides the geometers of his time for poor use of language. The language of the geometers is in the active mode, calling for extending lines and drawing circles. Plato would prefer the geometers to express themselves in the passive. For example, they should say: "there exists a line that intersects the circle" as opposed to "draw a line that intersects the circle"; Plato, *The Republic*, G. R. Ferrari (ed.), T. Griffith (trans.) (Cambridge: Cambridge University Press, 2000), Book 7: 527a, b.
7. Cesare Burali-Forti (1861–1931) was an assistant to Guiseppe Peano (1858–1932), a founder of set theory. See Haskell Curry, *Foundations of Mathematical Logic* (New York: McGraw-Hill, 1963), 5 n.2.
8. For an in-depth discussion see Marcus Giaquinto, *The Search for Certainty; A Philosophical Account of the Foundations of Mathematics* (Oxford: Clarendon Press, 2002). Ernst Zermelo (1871–1953), who set out to axiomatize set theory in 1908, did not develop a theory of classes; his was a theory of sets. As such, for Zermelo talk of proper classes lies outside mathematics, because it lies outside set theory. In particular, we cannot talk within set theory of the proper class of all the ordinals. Thus, we are never in a position to run the paradox. We shall return to these ideas later in the chapter.
9. We could marshal our vocabulary and say that a mathematical realist is a realist in truth-value, and that a platonist is a realist in both ontology and truth-value. This marshalling would help to distinguish the various realist positions. Unfortunately, such clarity would come at a price, and the price would be confusion when reading the literature. Unfortunately, in the literature, many realists do not draw the distinction between realism and platonism in this way.
10. Notice that we have been using "see", "perceive" and "intuit" interchangeably. These notions will be disentangled in the course of the book. For now, we lump them together.
11. Set theory was developed at the end of the nineteenth century and the beginning of the twentieth century. Work is still being done in set theory. The theory grows up around

the notion of a set of objects. We explore this notion, and discover the idea that there must be an empty set: a set containing no objects. We go on to discover other sets, we develop axioms which allow us to find more sets out of existing sets, and so on. The "universe of sets" thus discovered/created, is arranged in a hierarchy, and called "the set-theoretic hierarchy", or "the set-theoretic universe".

12. A well-formed formula is a string of symbols in the formal language that is grammatical. That is, it makes sense. "*P* & ~*P*" is a well-formed formula, but "~~*P*&" is not a well-formed formula.
13. At this stage, simply read this at face value and we shall assess it and discuss alternative views in depth later. We leave the logic in place for now, and question it only later, when we have a viable alternative.
14. Russell changed his mind frequently. For a careful and illuminating navigation through Russell's various theory changes see Michael Potter, *Reason's Nearest Kin: Philosophies of Arithmetic from Kant to Carnap* (Oxford: Oxford University Press, 2000); ch. 5, 119–63, is on Russell's type theory.
15. Maddy prefers the term "realism" over "platonism" (or "Platonism") to describe Gödel's position: Penelope Maddy, "Perception and Mathematical Intuition", in *The Philosophy of Mathematics*, W. D. Hart (ed.), 114–41 (Oxford: Oxford University Press, 1996), 114 n.1.
16. Very simply, this is because of the causal theory of knowledge is still philosophically popular, especially in philosophy of science. The immediate problem with a causal theory of knowledge and abstract objects is to account for how it is that abstract objects can have a causal connection with us at all.
17. Supervience is a relation between, usually, an object and a concept. We say that "the concept supervenes on the object". This means that were the object to change, then so would the concept. The clearest example is in aesthetics. We say that beauty supervenes on a painting. Should the painting be altered, so would the beauty. "Beauty" is abstract whereas the painting is physical.
18. The list of such mathematicians is too long to include in full. Examples include Albert Dragalin, Smolenski and Alexander Yessinin-Volpin. Some of the Polish logicians of the 1920s and 1930s also have a strong awareness of constructivism.
19. I highly recommend P. L. Heath, "Nothing", in *The Encyclopedia of Philosophy*, vol. 5, Paul Edwards (ed.), 524–5 (New York: Macmillan, 1967).
20. "Quite certain" means that we have proofs of equi-consistency. These are proofs that if one system is consistent, then so is the other. We do not have absolute proofs of consistency of any set theories. Since there are no absolute proofs of the consistency of any set theory, we ultimately have to express our faith that so much of our mathematics could not be wrong, since many of the set theories developed are equi-consistent with each other. We might then add that we have not discovered a paradox yet in the existing theories, so we are unlikely to. Of course, this is an empirical inductive argument.
21. This is enough to distance Köhler from Maddy.
22. For readers to whom this will make some sense: "rational intuition" is comparable to Kant's notions of spatial intuition and temporal intuition.
23. One person will have a better arithmetical sense than another if he is quicker or more precise in making calculations or estimates than another.
24. See the character called the "Gödelian optimist" in Neil Tennant, *The Taming of the True* (Oxford: Clarendon Press, 1997). This character surfaces in a few places in the

text. Köhler is a Gödelian optimist, as are the mathematicians he is representing in his philosophical position.
25. Of course, this judgement is made in reference to the few fragments of ancient Greek writing we have.
26. There are different stances towards a causal theory of knowledge. We can say that a causal account is the only possible account of knowledge, and "causal" is read as "physically causes", in which case we cannot have knowledge of abstract objects. Or we build into our theory of causation some notion of rationality, so our mathematical skills are part and parcel of our causal theory of knowledge. The third possibility is that we have two theories of knowledge – one for physical objects, one for abstract objects – and we try to use the appropriate one for the occasion.

Chapter 3. Logicism

1. In English, *Concept Script: A Formula Language, Modeled upon that of Arithmetic, for Pure Thought*, in *From Frege to Gödel: A Source Book in Mathematical Logic, 1879–1931*, J. van Heijenoort (ed.), 1–82 (Cambridge, MA: Harvard University Press).
2. In English, *The Foundations of Arithmetic*, J. L. Austin (trans.) (Evanston, IL: Northwestern University Press, 1980).
3. The *Grundgesetze* is partially translated in *Translations from the Philosophical Writings of Gottlob Frege*, 3rd edn, Peter Geach and Max Black (ed. and trans.), (Oxford: Blackwell, 1980), 117–224.
4. Frege changed his mind about this towards the end of his career. However, we shall restrict ourselves to the logicism he developed in the *Begriffsschrift*, *Grundlagen* and *Grundgesetze*. In the *Grundlagen* (see §13), Frege clearly states that geometry is not reducible to logic.
5. See my "A *Reductio Ad Absurdum* Argument for Naïve Logicism", unpublished manuscript.
6. George Boole, *An Investigation of the Laws of Thought* (New York: Barnes & Noble, 2005), 192–228.
7. Sometimes "propositional logic" is referred to as "sentential logic".
8. An alternative standard symbol for "→" is "⊃". Here, "→" is the symbol for implication; it is read as "if ... then ...".
9. Modern propositional logic was developed separately by Emil Post and Wittgenstein in 1920. Some medieval logicians also developed a version of propositional logic. See Martha Kneal and William Kneal, *The Development of Logic* (Oxford: Oxford University Press, 1962). Unfortunately, what the medieval logicians developed was not very popular, and the syllogistic logic predominated.
10. Dedekind, *Essays on the Theory of Numbers*.
11. That is, we might have thought that we had two different structures, because they were differently described, but we only have one that we can detect. The last qualifier is important. What we can detect depends on the sophistication of our mathematical language. The realist maintains that there is only one structure really, and this does not depend on our ability to detect it, or confound it with others. This is an interesting point of comparison with structuralism.
12. This is not really the place to raise the objections but, roughly, they have to do with the limitative results that pertain to full (as opposed to Henkin) second-order logic. Full

second-order logic is incomplete, not compact and does not have the Löwenheim–Skolem properties. This makes for a less "tidy" system, to use an aesthetic term, than first-order logic.
13. It is not true that every number is even. However, it is true that the conditional of 7' holds since the antecedent of the conditional is false. 0 is not an even number; nor is it the case that if x is an even number its immediate successor will always be an even number.
14. A set is isomorphic to another if and only if there is a one-to-one correspondence between the sets.
15. Ironically, (because compactness is supposed to be "good"), we can construct such a non-standard model by exploiting the compactness of first-order logic.
16. Some philosophers deny that there are any abstract objects. However, see Bob Hale, *Abstract Objects* (Oxford: Blackwell, 1987), which defends a view of abstract objects.
17. Put more carefully, any formal system that looks like an alternative had better be parasitic on regular arithmetic. For example, modular arithmetic, or clock arithmetic, is not a real alternative to regular full arithmetic. It is parasitic on regular arithmetic, for there is no principled (logical/mathematical) upper bound on the module. We can do arithmetic mod 8, or mod 9 or mod 10 and so on. One might think that first- and second-order arithmetic are alternatives to each other, and that one is true and the other not always true. I think, and this is speculation, that Frege would have referred to non-standard models of arithmetic. These are only constructible in first-order arithmetic. This indicates that second-order arithmetic is the more loyal formal representation of arithmetic.
18. For those unfamiliar with this vocabulary, here is the explanation. A conditional is an "if ... then ..." statement. Often we symbolize this with $\rightarrow$ or $\supset$. In propositional logic we might write $P \rightarrow Q$. P is the antecedent of the conditional and Q is the consequent. *Modus ponens* is the rule that says that if you have, for example, $P \rightarrow Q$, and independently of this you also have P, then you may write Q on a new line. That is, you may infer Q.
19. Frege is careful about this. The claim is not that someone with no sense experience at all can still come up with mathematics. Rather, recognizing arithmetic truths does not depend on any particular sense experience. Frege acknowledges that as human beings we probably need some sense experience to get us started in thinking at all. See *Grundlagen* §105, n. 2.
20. See, for example, Michael Dummett, "Frege and Kant on Geometry", *Inquiry* **25** (1980), 233–54.
21. For a nice discussion of this see Norma Goethe, "Frege Between Kant and Leibniz or How to Understand Truth by Means of Rigorous Proof ", manuscript presented to the History of Philosophy of Science Working Group (HOPOS), June 2002, and Eckehart Köhler, "Logic is Objective and Subjective", paper presented at the History of Philosophy of Science Working Group (HOPOS) Conference, Vienna, July 2000.
22. In particular, Frege agreed with Kant about geometry requiring spatial intuition.
23. Quoted in Stewart Shapiro, *Philosophy of Mathematics: Structure and Ontology* (Oxford: Oxford University Press, 1997), 144.
24. Ernst Zermelo (1871–1953) had discovered the paradox before Russell, but not in Frege's system. Zermelo discovered it in an early attempt to axiomatize set theory, and he knew to avoid the problem. Russell made it famous because of the dramatic circumstances under which he revealed it.

25. Peter Geach and Max Black, "Frege on Russell's Paradox (appendix to Vol. II)", in *Translations from the Philosophical Writings of Gottlob Frege*, P. T. Geach (trans.) (Oxford: Blackwell, 1952), 214.
26. For a discussion of intensional logic see Edward N. Zalta, *Intensional Logic and the Metaphysics of Intentionality* (Cambridge, MA: MIT Press, 1988).
27. Set theory usually has a comprehension principle or a principle of extensionality. Restrictions are placed on this in order to avoid paradox, but it is as though mathematicians regret the fact that the naive version of the principle leads to contradiction.
28. Frege even allows us to think of contradictory notions, such as all objects not equal to themselves. Since this is contradictory, no object falls under it. He uses this to pick out the number 0, or the empty set. Frege notes that any notion that has no objects falling under it will do, so we could also use "state in Canada" to pick out 0, since there are none. By basic law V, the number of "state in Canada" and the number of "object not identical with itself" are identical since they are equivalent in extension.
29. For a recent discussion of the Julius Caesar problem see Bob Hale and Crispin Wright, "To Bury Caesar ...", in *The Reason's Proper Study: Essays Towards a Neo-Fregean Philosophy of Mathematics*, B. Hale & C. Wright (eds), 335–96 (Oxford: Oxford University Press, 2001).
30. There is some controversy concerning the order in which to write the names Russell and Whitehead. It is clear, from their other writings and correspondence that Whitehead's contribution was the more technical aspects of the work, and Russell contributed more to the philosophical aspects. Many logicians consider that the technical achievement is the greater of the two, and therefore favour writing Whitehead's name before Russell's, and Whitehead is the first named author on their *Principia Mathematica* (1910–13).
31. Frege also thought that he could probably reduce analysis to logic as well, but he was also clear that he did not think that geometry was reducible to logic. In contrast, Whitehead and Russell do think that geometry is reducible to their formal system of logic.
32. Here, "intuitively" is meant in the sense of pre-theoretic, or informal.
33. "Theorems" are the syntactic counterparts of "truths". Note too that in studying Frege carefully one should be aware that the semantic–syntactic distinction is not as explicit and natural as it is now, so language that is sensitive to the semantic–syntactic distinction is somewhat anachronistic. The distinction only became significant after Gödel's incompleteness results were understood by the mathematical community.
34. An interesting way of thinking about this is that the grammar rules and axioms implicitly bind the quantifiers. In particular, they bind the universal quantifier. That is, "all" does not mean "anything that we can think of out of the blue". Instead, the universal quantifier is bound by the hierarchical structure. Concepts of different types have to be built from the ground up: we have to start with 0 and add grammatical structure to that piecewise. This notion of building from the ground up, and building piecewise, is an important anti-realist concept.
35. The existential and universal quantifiers are interdefinable so, strictly speaking, it does not matter whether we are discussing the universal, the existential or both. The existential is definable as the negation of the universal not. We can give an example expression: $\exists x(Fx)$ is definitionally equivalent to $\sim\forall x(\sim Fx)$. However, intuitively, it is the universal that generates paradoxes.
36. More precisely, if Zermelo–Fraenkel set theory is consistent, then type theory is consistent. This is called "relative" consistency, as opposed to absolute consistency. The

latter is where we have proof that we shall not generate contradiction from a set of axioms. Relative consistency is the best sort of result we can get for formal systems of a certain minimal complexity (such as that of first-order arithmetic). This was shown by Gödel in 1931.
37. See Stewart Shapiro, *Thinking About Mathematics* (Oxford: Oxford University Press, 2000), 120–21.
38. See Crispin Wright, *Frege's Conception of Numbers as Objects* (Aberdeen: Aberdeen University Press, 1983).
39. More impressive still, we can recapture basic set theory too, by adding another principle, which is similar in structure to the numbers principle.
40. Witness the title of the latest book-sized defence of this sort of logicism: Bob Hale and Crispin Wright (eds), *The Reason's Proper Study: Essays Towards a Neo-Fregean Philosophy of Mathematics* (Oxford: Oxford University Press, 2001).
41. Filling in the logic of the argument in detail, grammatically the numbers principle is a declarative sentence, in the sense that it can take on a truth-value (unlike a question, for example). Moreover, a declarative sentence cannot be both analytic and synthetic. Furthermore, a declarative sentence has to be one or the other.
42. The full argument also has to interpret the universal quantifiers ranging over the whole expression. Be aware that this is a simplification of the full argument. For the full argument see Wright, *Frege's Conception of Numbers as Objects*, and further elaborations on the argument given by Hale and Wright.
43. The article first appeared in Richard Heck Jr (ed.), *Language, Thought, and Logic* (Oxford: Oxford University Press, 1997), ch. 9.
44. Köhler does not specifically address the numbers principle in his paper, so this is an extension of his position that he might not accept.
45. There is much material on this. See the Guide to Further Reading.
46. There are non-classical developments of logicism. In *Autologic* (Edinburgh: Edinburgh University Press, 1992), Neil Tennant arguably develops an intuitionist version of logicism. He argues that a relevant intuitionist logic is fundamental as normative of good reasoning, and then reduces arithmetic to this. We shall examine this further in Chapter 5.

Chapter 4. Structuralism

1. A "function" (sometimes called a "mapping" or "graph") takes us from one set of mathematical objects to another. An example of a function is "add two". This takes us from a set of numbers, the "domain", to another set of numbers, the "range". The domain is independent of the function; we can specify whatever we like to be the domain. We then carry out the function, and get the range.
2. Ernie is an oblique reference to Ernst Zermelo, and Johnny is an oblique reference to Johann von Neumann. Interestingly, Benacerraf reverses the two when he identifies how each develops the set-theoretic analogue of the pre-set theoretic numbers.
3. More technically: the Peano/Dedekind axioms are true of the Zermelo numerals, and the finite ordinals of von Neumann.
4. In Zermelo–Fraenkel set theory the ordinals (and therefore, finite cardinals) are "constructed" *ex nihilo*, that is, from the empty set, as follows. We symbolize 0 by $\emptyset$. Then 1 is symbolized by the set of the empty set: $\{\emptyset\}$. Now 2 is the set whose members are the

empty set and the set of the empty set: {∅, {∅}}. And 3 is the set whose members are 0, 1 and 2: the empty set, the set of the empty set and the set of the empty set together with the set of the empty set – {∅, {∅}, {∅, {∅}}}. In contrast, in von Neumann set theory the numbers are constructed *ex nihilo* again from the empty set, but the number depends on the number of set-theoretic brackets, so 0 is ∅, 1 is {∅}, 2 is {{∅}}, 3 is {{{∅}}} and so on.

5. Note that this is always a risk with platonism or realism. The realist position includes in it the possibility that we are wholly wrong in thinking that we track, accurately, *reality*. Benacerraf's puzzle makes it plain that we have got it wrong, at least some of the time, and that there is no way to work out how to ensure that we get it right.
6. This happens when we are reasoning over choices. For example, I might be trying to choose between studying for a test or going to a party. Both are possibilities available to me. I might reason as follows. If I go to the party and do not study, then there is a possibility that I will fail the test. If I were to fail the test, then that would be disastrous, so the very possibility of failing had better be eliminated by my studying. Reasoning over probabilities is a refinement on this sort of example.
7. In modal logic, we distinguish between a possible world (possible structure for Hellman) and the universe of possible worlds. We cannot talk of the structure of all structures/world of all possible worlds, but just the universe of structures/possible worlds. The universe is a logic: a set of rules governing inferences concerning the relationships between the worlds.
8. In *Mathematics Without Numbers: Towards a Modal-Structural Interpretation* (Oxford: Clarendon Press, 1989), Hellman assumes set theory as his background theory governing what is, and is not, possible. However, in a more recent article, "Structuralism", in *The Oxford Handbook of Philosophy of Mathematics and Logic*, S. Shapiro (ed.), 556–62 (Oxford: Oxford University Press, 2005), Hellman gives the outlines of a modal structuralism based on second-order logic. So choosing set theory as a backdrop is not a necessary move.
9. Here we mean "objects" as traditionally conceived. Although this can have two readings, both are acceptable. The first is the realist reading, where the objects of mathematics are the ontology of mathematics. These are the things about which one develops a mathematical theory. Examples of mathematical objects are numbers, points, lines and fractions. The other reading is that the objects of mathematics are whatever it is that a first-order mathematical theory has in its intended domain. So, for example, first-order arithmetic has the natural numbers in the domain of quantification. The restriction to "first-order" is important, since we do not want to talk of relations as objects. In the last sentence, the word "object" slipped up an order, and we want to restrict our use of object to the lowest level, for now. However, the structuralist will be asked later whether the relations in a structure are objects of mathematics.
10. If one is willing to give up this intuition, then one is inclined towards constructivism.
11. I am not sure I want to invoke the supervenience relation here. For those unaware of the meaning of "supervenience" it has several meanings or formulations, but the general idea is sufficient here. A property (structure) supervenes on an object (set of objects) just in case some changes in the object (set of objects) will effect a change in the property (structure). A famous example is that the beauty of a painting supervenes on the brushstrokes of the paint on the canvas. The material of the painting is insufficient to make the painting beautiful, and some alterations in the arrangement

of paint may not affect the beauty, but some rearrangements will, as will cutting up the canvas and distributing the parts.
12. Study of the number 8 in isolation from the other numbers would approach numerology, or some spiritual study of number as entity or force. This is not the sort of study engaged in by mathematicians, despite there being whole books devoted to the study of π. In fact, π is studied as a member of the irrational numbers, and because it figures in some geometry theorems.
13. In the quotation, the emphasis on effectiveness is acknowledged to be a matter of choice. The notion of structure is, however, viewed as central to mathematical study generally.
14. Since we do not have Babylonian proofs or general methods, we can surmise that the tables were filled in as much by approximation as by use of a general technique.
15. Of course, someone had to know something general in order to generate a table in the first place. However, the method of table creation was not widely known; at least we have no evidence that it was. The tables are incomplete and surprisingly, but not completely, accurate.
16. The Löwenheim–Skolem theorem states that if T is a countable theory having a model, then T has a countable model; Joseph R. Shoenfield, *Mathematical Logic* (Nantick, MA: A.K. Peters, 1994), 79. A theory is countable just in case the language in which the theory is expressed has no more than countably many non-logical constants. "Countable" means $\aleph_0$ or lower (*ibid*.: 78).

   The compactness theorem of a mathematical theory is: "A formula **A** in a theory T is valid in T iff it is valid in some finitely axiomatized part of T". A corollary to this is: "A theory T has a model iff every finitely axiomatized part of T has a model" (*ibid*.: 69).

   There are results about "degrees of compactness" where, rather than specify that the theory be finitely axiomatized, there is some infinite cardinal, less than which is the number of axioms needed to show validity. Specifically, let κ be some infinite cardinal. Then for κ-compactness, a formula **A** in a theory T is valid in T iff it is valid in some sub-version of T with only κ axioms.
17. The logicist will tend to insist that logic is even more general than the rest of mathematics. This is important if we exploit some sort of hierarchy of knowledge in our philosophy.
18. For the philosophical sophisticates, we are discussing ontological reduction, as opposed to epistemic, justificatory or explanatory reduction. Often all the different forms of reduction are conflated.
19. Other structuralists, such as Hellman, advocate a modal structuralism. They relegate the ontological question to possible worlds. That is, Hellman talks of possible and necessary structures. We then ask what it is that makes a structure necessary, as opposed to possible. That will depend on the modal logic used (which axioms the logic has); Hellman prefers S5 (*Mathematics Without Numbers*, 17 n. 8).
20. The reason for disallowing this sort of question is that it could lead to paradox, which is always a danger with unrestricted quantification.
21. A very interesting question has to do with the structure of the real numbers, and how big it is: what the cardinality of the real numbers is. The cardinality of the real numbers depends on which structure we are measuring them from. This is known as the Skolem paradox. For a nice presentation of Skolem's paradox see the appendix in Moshe Machover, *Set Theory, Logic and their Limitations* (Cambridge: Cambridge University Press, 1996).

22. An example of this came up at a talk in the logic colloquium at George Washington University. A well established model theorist, Valentina Harizanov, remarked about some mathematical concept being discussed that "this property is important [mathematically], but it is not recognized in model theory" (unpublished).
23. See Michèle Friend "Meinongian Structuralism", in *The Logica Yearbook 2005*, M. Bílková and O. Tomala (eds) (Prague: Filosofia, 2006).

Chapter 5. Constructivism

1. A "prime pair" is a pair of primes separated by only one (even) number. Examples of prime pairs are: <3, 5>, <5, 7>, <11, 13>, <15, 17>, ...
2. The philosophical language is at odds with ordinary discourse where we might say that someone is realistic, or a realist if she refuses to talk of verification-transcendent truths. A philosopher would label such a person an anti-realist. The philosophical use of the term relates to our notion of reality, and whether this depends on us or is independent of us.
3. Philosophers taught in the United States tend to draw the distinction between a realist and an anti-realist somewhat differently. For an American, a relativist is an anti-realist; the rest are largely realists. The way in which the distinction is drawn here, and in Chapter 2 is more common to the UK. Canadians tend to know of both distinctions. Students should be aware of the differences in order to make sense of the literature on these subjects. For a discussion of the differences see Tennant: *The Taming of the True*, 4–6.
4. These are all taken from Michael Dummett, *Elements of Intuitionism*, 2nd ed. (Oxford: Oxford University Press, 2000), 89.
5. Soundness and completeness refer to the match between semantic proofs and syntactic proofs. A formal system is sound if every syntactic rule is truth-preserving. A formal system is complete if every semantic truth has a corresponding syntactic proof. Not all formal systems are complete.
6. Recall the discussion in Chapter 3 about Frege's notion of "basic law", and how that contrasts to notions of "axiom".
7. This is a nice point of contact between the constructivist and the formalist, who we shall encounter in Chapter 6.
8. In many texts introducing logic to students, the complexity of disjunction is avoided by giving the student a rule called "disjunctive syllogism": $A \vee B, \sim A \vdash B$. This is provable from disjunction elimination.
9. De Morgan's laws are very practical. They were developed by Augustus De Morgan in the 1920s. They are: $\sim(A \wedge B) \vdash \sim A \vee \sim B$ and $\sim(A \vee B) \vdash \sim A \wedge \sim B$. The inferences work in the other direction too, in classical logic.
10. Tautologies are semantic and theorems are syntactic. Tautologies are always true. Theorems are proved from no premises using natural deduction. Assumptions are all discharged. If a formal system is complete then exactly the same well-formed formulas will be theorems and tautologies.
11. For good discussions about this see Stephen Read, *Thinking About Logic: An Introduction to the Philosophy of Logic* (Oxford: Oxford University Press, 1994), 59 and Graham Priest, *An Introduction to Non-Classical Logic* (Cambridge: Cambridge University Press, 2001), 151.

12. For a nice discussion of why this was, placed in a historical context, see Paolo Mancosu, *From Bouwer to Hilbert: The Debate on the Foundations of Mathematics in the 1920s* (Oxford: Oxford University Press, 1998).
13. For those interested in other responses, where we block the reasoning minimally, I suggest looking at Russell's *The Principles of Mathematics* (Cambridge: Cambridge University Press, 1903) for the Russell paradox. See also R. M. Sainsbury, *Paradoxes* (Cambridge: Cambridge University Press, 1988). Also, for a very close and original analysis of Russell's paradox in Frege, with a sensitivity to constructivism, see Alan Weir, "Naïve Set Theory is Innocent!", *Mind* **107** (1998), 763–98.
14. This is related, or is a species of, reverse mathematics. Reverse mathematics was proposed by Harvey Freedman and others in the 1980s. The idea is to show which axioms a theorem rests on, for often the first proof of a theorem uses more axioms than it needs to. Reverse mathematics tries to minimize the number of axioms needed. Constructivists, such as Douglas Bridges or Errett Bishop, do something similar. They are only interested in constructively acceptable axioms and rules of inference. Bridges and Bishop are engaged in a project of going through a number of important proofs and rewriting them in constructively acceptable terms. It is not clear, at this stage, how much of classical mathematics they can recapture.
15. To put it another way, they give us more information that we find interesting. Maybe ultimately this is just a preference, and in that sense a matter of taste.
16. It is interesting to compare this to formalism.
17. This should remind us of Aristotle and the notion of the potential infinite.
18. "Holds" cannot mean "is provable" because of the decision problem that is, it is a well-established fact in mathematics that there are some formulas that cannot be proved syntactically. However, we might be able to use a semantic argument to show that they are true or false.
19. In propositional logic, this will be an argument that shows that every truth-value assignment to the proposition letters that makes the premises true, also makes the conclusion true.
20. The full axiom of choice should be added to this list. It was excluded simply because it was not woven into the interconnection of ideas previously given. The axiom of choice is also intimately connected to the law of excluded middle, but to show this requires the introduction of technical vocabulary. For a good exposition of these intricacies, see W. Tait, "The Law of Excluded Middle and the Axiom of Choice", in his *Essays in the Philosophy of Mathematics and its History*, 105–32 (Oxford: Oxford University Press, 2005).
21. An example of appropriate background assumptions would be if A is finitely checkable. If this is the case, then we can turn the doubly negated A into a positive A, not by means of two *reductio* proofs, but by means of a positive proof going through the examples.
22. A partial-order is like a tree. There is a, or several, base case(s). There are sentences of one degree of complexity greater, which are "derived from" or "based in" the base cases. There are sentences of one degree of complexity higher than those occupying level one, and derived from those on level one. For example, "Quiet", "Quiet, please" and "Quiet in here" might be thought of as occupying the first three levels, respectively. It is difficult, but not impossible, to give a detailed articulation for partially ordering sentences in terms of complexity. It is easier if we consider sentences written in a formal language, so proposition letters are the least complex, proposition letters together with the negation come next, then come pairs of proposition letters with a binary connective between, and so on.

23. The original quote is from David Hilbert, "Neubegründung der Mathematik", *Abhandlungen aus dem mathematischen Seminar der Hamburgischen Universität* **I** (1922), 200.

Chapter 6. A *pot-pourri* of philosophies of mathematics

1. This is not quite right. There might be other constraints as well, such as not being a "silly" system: one that is consistent but really quite useless. For example, we might have a very small system that can only prove one thing. Or we might have a system that can only prove falsehoods. Nevertheless, what is not a constraint on counting something as a mathematical system is "truth".
2. Stricker is cited in Frege, *The Foundations of Arithmetic*, v.
3. Referred to in Frege's *Grundlagen*, *The Foundations of Arithmetic*, §7.
4. Philosophers are often charged with doing something that is not useful.
5. Moosbrugger is a character in Robert Musil, *The Man Without Qualities* (New York: Coward-McCann, 1953).
6. Consider the computer games where several people participate with virtual characters. The entire game is like a work of fiction, but there are many players (authors).
7. That Stricker is one of the first to propose this is simply inferred from Frege's *Grundlagen*, where Frege heralds him as the champion of psychologism. I have not looked into the issue of the origin of the position.
8. See Claire Ortiz Hill and Guillermo E. Rossado Haddock (eds), *Husserl or Frege? Meaning, Objectivity and Mathematics* (Chicago, IL: Open Court, 2000).
9. This will come as a surprise to many philosophers brought up in the analytic tradition. However, see Hill and Haddock, *Husserl or Frege*, xiii–xiv. Richard Tieszen, *Phenomenology, Logic and the Philosophy of Mathematics* (Cambridge: Cambridge University Press, 2005) repeatedly points out that Frege and Husserl were both very critical of psychologism in the philosophy of mathematics, and their arguments against psychologism were very similar. For this reason Frege made a mistake in accusing Husserl of psychologism.
10. See, for example, Claire Hill, "Frege's Attack on Husserl", in Hill and Haddock (eds), *Husserl or Frege?*, 95–108, esp. 103.
11. This includes calculations concerning finite numbers, for example: adding, multiplying, subtracting and so on.
12. We can use an abacus or primitive calculator to perform this sort of arithmetic.
13. See, for example Paul Tomassi, *Logic* (London: Routledge, 1999) or E. J. Lemmon, *Beginning Logic* (Indianapolis, IN: Hackett, 1965).
14. The reasons why Tomassi, *Logic*, and Lemmon, *Beginning Logic*, begin with the natural-deduction rules are twofold: pedagogical and philosophical. Pedagogically, the motivation is simply that students who are first taught the truth-tables tend to depend on them too much when they then learn natural deduction, so they have difficulty understanding the rules since they are tempted to try to understand them in terms of the truth-tables. The philosophical reason derives from a sensitivity towards constructivist positions in the philosophy of mathematics. Indeed, rather than simply present an expedient set of rules of deduction, both texts are careful to give both introduction and elimination rules, thus preserving the symmetry requirement of the constructivist on rules of inference. This was discussed in Chapter 5.
15. Curry distances himself from Hilbert in not insisting on consistency of a mathematical theory as a constraint on theories.

16. Shapiro is quoting David Hilbert, "Über das Unendliche", *Mathematische Annalen* **95** (1925), 171.
17. Whether Hilbert is wholly successful in his demonstration is controversial. He certainly used intuitive ideas as a teaching device, but the idea was to do away with these and replace them with an understanding of the axioms and manipulation rules. If the rules and axioms are good, then intuition is left behind; Hilbert, *Foundations of Geometry*, 2nd edn, L. Unger (trans.) (La Salle, IL: Open Court, 1971).
18. To distance ourselves from notions of semantics as "giving meaning" we sometimes say that a model "satisfies" a theory. Many different models might satisfy the same theory, so a model does not give a unique meaning. A set of models satisfies a theory, so meaning, in the sense of satisfaction, is at best ambiguous.
19. This might be partly what led Russell astray, for in English, at least, we tend to talk about "being an object", so the existence of objects is built into the grammar. Russell was rather keen to model logic on grammar, albeit a universal grammar, and not English grammar. Nevertheless, English grammar was that with which he was most familiar. Russell criticized Meinong for allowing too much into his ontology, thus assuming that calling something an object is enough to give it ontological status: enough for it to exist, in some sense. If we allow fictional objects to exist, then we have many objects. If we allow contradictory objects to exist, then the contradiction spreads to infect everything and any theory becomes trivial, in the sense of inconsistent.
20. When a sentence can have both truth-values, this is called a "truth-value glut". A paraconsistent logic allows truth-value gluts.
21. The famous example Lakatos exploits to make this point is that of Euler's formula for a polyhedra, $V - E + F = 2$, where V is the number of vertices, E is the number of edges, and F is the number of faces. There are exceptions. For example, an open cylinder has two edges and one face. The definition of "vertex" is not fixed. It comes under discussion in Imre Lakatos, *Proofs and Refutations: The Logic of Mathematical Discovery*, J. Worrall and E. Zahar (eds) (Cambridge: Cambridge University Press, 1997), 107, 114–15. To make Euclid's formula work, there would have to be three vertices, and this is a bit odd under any intuitive understanding of "vertex". What Lakatos finds interesting is that there are several possible reactions to the ambiguity in the notion of vertex.
22. Lakatos classed counter-examples into two sorts: heuristic and logical. Heuristic counter-examples foster a side investigation, not directly affecting the original conjecture for which it is a counter-example. Logical counter-examples force one to give something up in the original conjecture. To muddy the waters, it is not always clear which sort a counter-example is. It depends to some extent on how the mathematical community reacts to it. See Brendan Larvor, *Lakatos: An Introduction* (London: Routledge, 1998), 15.
23. It is interesting to compare these remarks to those of Plato admonishing the geometers for using the active mode over the passive mode. Plato thought of mathematics as static (*Republic*, bk VII, lines 527a, b).
24. This is meant neither in the technical sense of "game theory" nor in the strict sense of "game" as the formalist uses the term. Instead, here, this is meant in a very loose sense of play.

Guide to further reading

Introduction

After reading this book one's next step should be to consult one of the following: Stephen Körner, *The Philosophy of Mathematics: An Introductory Essay* (New York: Harper & Row, 1962); Stewart Shapiro, *Thinking About Mathematics* (Oxford: Oxford University Press, 2000); James Brown: *Philosophy of Mathematics: An Introduction to the World of Proofs and Pictures* (London: Routledge, 1999). These are general texts that examine most of the issues covered in this book and do not present one particular philosophical position, but many. They are not aimed exclusively at the specialist and so are reasonably accessible to the beginner. Körner's book discusses Platonism, logicism, formalism and constructivism. It is nicely set out, giving the position, and then a lengthy criticism of it. Shapiro's book gives a more current treatment of the positions covered in Körner, including a chapter on structuralism. Brown's approach is interesting, for he does not have chapters just on the various positions, but also on themes, such as definitions, or applied mathematics. Students new to the field will find these books the most approachable, and they serve as good comparisons to this book.

Somewhat more difficult but equally broad in approach is Marcus Giaquinto, *The Search for Certainty: A Philosophical Account of the Foundations of Mathematics* (Oxford: Clarendon Press, 2002), which explores the crisis in the foundations of mathematics caused by the set-theoretic paradoxes and discusses various philosophical and mathematical reactions to those paradoxes. Two collections of essays that have long been standard research texts in the philosophy of mathematics are Paul Benacerraf and Hilary Putnam (eds), *Philosophy of Mathematics: Selected Readings*, 2nd edn (Cambridge: Cambridge University Press, 1983) and Jean van Heijenoort (ed.) *From Frege to Gödel: Mathematical Logic, 1879–1931* (Cambridge, MA: Harvard University Press, 1977). Both are heavily used resources, comprising selections of classic papers in the philosophy of mathematics that have for the most part appeared in journals. The books are not as accessible to the beginner, but are excellent for deepening one's knowledge of the various positions. In addition, Stewart Shapiro (ed.), *The Oxford Handbook of Philosophy of Mathematics and Logic* (Oxford: Oxford University Press, 2005) is a more recent collection of papers by current philosophers of mathematics, outlining and critiquing particular positions.

It is also worth consulting encyclopedias. Entries in *The Routledge Encyclopedia of Philosophy*, the older *The Encyclopedia of Philosophy* (New York: Macmillan, 1967) and the two internet encyclopedias, "Stanford Encyclopedia of Philosophy" (http://plato.stanford.

edu/) and the "Internet Encyclopedia of Philosophy" (www.iep.utm.edu) provide excellent further threads to follow via their bibliographies.

Chapter 1. Infinity

For Zeno, Gregory Vlastos, "Zeno of Elea", in *The Encyclopedia of Philosophy*, vol. 8, 369–79, is a succinct and reliable article on Zeno. It discusses the paradoxes of motion in the context of arguments against the possibility of motion, and also discusses arguments that Zeno makes concerning plurality. It also sets Zeno's work in context with his contemporaries.

For the post-Cantorian theory of infinite ordinals and cardinals, Moshe Machover, *Set Theory, Logic and their Limitations* (Cambridge: Cambridge University Press, 1996), is a good book for the more logically or technically minded. It is a very clear presentation of set theory aimed at upper-level philosophy students. There are separate chapters on ordinals and cardinals. It is difficult to read these chapters without either reading the previous three chapters, or having a fairly solid foundation in logic. On a more amusing note, see Raymond Smullyan, *What is the Name of this Book? The Riddle of Dracula and Other Logical Puzzles* (Harmondsworth: Penguin, 1990). This is a series of logical puzzles, starting with those of the knights and knave variety (where knights only tell the truth and knaves always lie). The puzzles become increasingly difficult, and some of them require "diagonalization" to solve them.

For good philosophical overviews about infinity see A. W. Moore, *The Infinite* (London: Routledge, 1990) and J. R. Lucas, *The Conceptual Roots of Mathematics* (London: Routledge, 2000).

Chapter 2. Mathematical Platonism and realism

For early Platonism, see Plato's *Meno* and *Theatetus*. These can be readily found in many collections of Plato's dialogues. Both explore the peculiarities of mathematical knowledge over other sorts of knowledge based on sense experience. The claim defended in the dialogues is that mathematical knowledge is not taught, but is available to all reasoning people. For Plato's ontological views see the *Republic*. For general overviews on mathematical Platonism and realism see Körner, *The Philosophy of Mathematics*; Mark Balaguer, *Platonism and Anti-Platonism in Mathematics* (Oxford: Oxford University Press, 1998); Shapiro, *Thinking About Mathematics* and Brown, *Philosophy of Mathematics*.

For a good treatment of how and why the set-theoretic paradoxes caused a crisis in the philosophy of mathematics see Giaquinto, *The Search for Certainty*. For other developed views about set theory and its relation to philosophy see Michael Potter, *Set Theory and its Philosophy* (Oxford: Oxford University Press, 2004). For more specific views see Kurt Gödel, "What is Cantor's Continuum Problem?", in Benacerraf & Putnam (eds), *Philosophy of Mathematics*, 483–4, and "Russell's Mathematical Logic", in Benacerraf and Putnam (eds), *Philosophy of Mathematics*, 447–69. For Maddy, see Penelope Maddy, *Realism in Mathematics* (Oxford: Clarendon Press, 1990) and *Naturalism in Mathematics* (Oxford: Clarendon Press, 1997). Köhler's papers are in manuscript form.

Chapter 3. Logicism

The reading in this area is quite technical and can sometimes be quite challenging. For Frege one should really read the *Grundlagen* [*Die Grundlagen der Arithmetik*], *The Foundations of Arithmetic*, 2nd rev. edn, J. L. Austin (trans.) (Evanston, IL: Northwestern University Press, 1980). This is one of the classic texts in analytical philosophy. Furthermore, it is short and even entertaining in places. For Whitehead and Russell, the original is, of course *Principia Mathematica*, 3 vols (Cambridge: Cambridge University Press, 1910–13). This is not easy to understand, so the secondary literature is helpful. See, for example, the relevant chapters in Michael Potter, *Reason's Nearest Kin: Philosophies of Arithmetic from Kant to Carnap* (Oxford: Oxford University Press, 2000) and Giaquinto, *The Search for Certainty*.

To read another good exposition of Frege, and then Wright's development of Frege, read Crispin Wright, *Frege's Conception of Numbers as Objects* (Aberdeen: Aberdeen University Press, 1983). For more recent work on the neo-Fregean position as developed by Wright and Hale, see B. Hale and C. Wright, *The Reason's Proper Study: Essays Towards a Neo-Fregean Philosophy of Mathematics* (Oxford: Clarendon Press, 2003). For the "bad company argument" against Hale and Wright, see George Boolos, "The Standard of Equality of Numbers", in his *Logic, Logic, and Logic*, 214–15 and "Is Hume's Principle Analytic?", in *Logic, Logic, and Logic*, 301–14 (Cambridge, MA: Harvard University Press, 1998). For Köhler's views see his "Gödel on Intuition, and How Carnap Abandoned Empiricism" (unpublished manuscript) and "Logic is Objective *and* Subjective", paper presented at the History of Philosophy of Science Working Group (HOPOS) Conference, Vienna, July 2000. For Neil Tennant's development of logicism from an intuitionist perspective, see his *Anti-Realism and Logic: Truth as Eternal* (Oxford: Oxford University Press, 1987).

Chapter 4. Structuralism

For Benacerraf's views see two papers in Benacerraf & Putnam (eds) *Philosophy of Mathematics*: "What Numbers Could not Be", 272–94, and "Mathematical Truth", 403–20. For Hellman, see his *Mathematics Without Numbers: Towards a Modal-Structural Interpretation* (Oxford: Clarendon Press, 1989). For Resnik's position see his *Mathematics as a Science of Patterns* (Oxford: Clarendon Press, 1997) and his papers: "Mathematics as a Science of Patterns: Ontology and Reference", *Noûs* **15** (1981), 529–50, and "Mathematics as a Science of Patterns: Epistemology", *Noûs* **16** (1982), 95–105. For Shapiro, see *Philosophy of Mathematics: Structure and Ontology* (Oxford: Oxford University Press, 1997).

Chapter 5. Constructivism

For Brouwer, a less-known but quite readable article appears in the *Proceedings of the Irish Academy*: L. E. J. Brouwer, "The Effect of Intuitionism on Classical Algebra of Logic", *Proceedings of the Royal Irish Academy*, vol. 57, Section A: Mathematical Astronomical, and Physical Science, 113–16 (Dublin: Hodges, Figgis & Co. 1954–56). The paper is part of a centenary celebration of the publication of Boole's *Laws of Thought*. It is odd to find the paper here, since Brouwer pays little attention to Boole's work. Instead he gives an, uncharacteristically, clear and concise articulation of his philosophical views. For more on Brouwer see Paolo Mancosu: *From Brouwer to Hilbert: The Debate on the Foundations of Mathematics in the 1920s* (Oxford: Oxford University Press, 1998). For a more philosophically oriented discussion see Michael Dummett, *The Logical Basis of Metaphysics* (Cambridge, MA: Harvard University Press, 1991), and *Elements of Intuitionism*, 2nd edn (Oxford: Oxford University Press, 2000). The last text is good for an in-depth discussion of what can and what cannot be proved intuitionistically. For an excellent discussion of the manifestation argument there are several chapters and sections in Neil Tennant, *The Taming of the True* (Oxford: Clarendon Press, 1997).

Chapter 6. A *pot-pourri* of philosophies of mathematics

For more on empiricism see John Skorupski, "Later Empiricism and Logical Positivism", in *The Oxford Handbook of Philosophy of Mathematics*, 51–74. See also John Stuart Mill, *A System of Logic, Ratiocinative and Inductive* (London: Longman, 1970). For fictionalism see Hartry Field: *Science Without Numbers: A Defence of Nominalism* (Oxford: Blackwell, 1980). There is also a nice short dicussion of fictionalism in his *Realism, Mathematics and Modality* (Oxford: Blackwell, 1989). For a thorough discussion of the indispensability argument, see Mark Colyvan, *The Indispensability of Mathematics* (Oxford: Oxford University Press, 2001).

There are no readily available sources on psychologism. General treatments can be found in encyclopedias. Husserl is discussed nicely in Richard Tieszen, *Phenomenology, Logic and the Philosophy of Mathematics* (Cambridge: Cambridge University Press, 2005).

Formalism is described in Curry's work. There is not much on Curry so it is best to look at his own books. The introductions and notes are philosophically explicit and candid. See Haskell B. Curry, *Foundations of Mathematical Logic* (New York: McGraw-Hill, 1963), "Remarks on the Definition and Nature of Mathematics", in Benacerraf & Putnam (eds), *Philosophy of Mathematics*, 202–6, and especially *Outline of a Formalist Philosophy of Mathematics* (Amsterdam: North Holland, 1951).

For more on Hilbert, see Michael Hallett, "Physicalism, Redutionism and Hilbert", in *Physicalism in Mathematics*, A. D. Irvine (ed.), 182–256 (Dordrecht: Kluwer, 1989), Michael Hallett, "Hilbert, David", in *Handbook of Metaphysics and Ontology*, H. Burkhardt & B. Smith (eds), 354–8 (Munich: Philosophia, 1991) and Michael Detlefsen, *Hilbert's Program* (Dordrecht: Kluwer, 1986). To read some of the original, see David Hilbert, *Foundations of Geometry*, 2nd edn, L. Unger (trans.) (La Salle, IL: Open Court, 1971). See also the relevant sections in Mancosu, *From Brouwer to Hilbert*; Shapiro, *Thinking About Mathematics*; Brown, *Philosophy of Mathematics* and Potter, *Reason's Nearest Kin*.

For more reading on Meinongian mathematics see Richard Routley, *Exploring Meinong's Jungle and Beyond* (Canberra: RSSS, Australian National University, 1980); Richard Sylvan, "The Importance of Nonexistent Objects and of Intensionality in Mathematics", *Philosophia Mathematica* 11 (March 2003), 20–52; Graham Priest, *In Contradiction* (Dordrecht: Kluwer, 1987) and "Meinongianism and the Philosophy of Mathematics", *Philosophia Mathematica* 11 (February 2003), 3–15.

For the original Lakatos see Imre Lakatos, *Proofs and Refutations: The Logic of Mathematical Discovery*, J. Worrall & E. Zahar (eds) (Cambridge: Cambridge University Press, 1997). For accessible secondary reading see Brendan Larvor, *Lakatos: An Introduction*, (London: Routledge, 1998). For deeper scholarly issues surrounding Lakatos see George Kampis, Ladislav Kvasz and Michael Stöltzner (eds) *Appraising Lakatos: Mathematics, Methodology and the Man* (Dordrecht: Kluwer, 2002).

Bibliography

Anglin, W. S. & J. Lambeck 1995. *The Heritage of Thales*. New York: Springer.
Anonymous 1994. *The Epic of Gilgamesh*, S. Shabandar (trans.). Reading: Garnet Publishing.
Balaguer, M. 1998. *Platonism and Anti-Platonism in Mathematics*. Oxford: Oxford University Press.
Benacerraf, P. 1983a. "What Numbers Could Not Be". See Benacerraf & Putnam (eds) (1983), 272–94.
Benacerraf, P. 1983b. "Mathematical Truth". See Benacerraf & Putnam (eds) (1983), 403–20.
Benacerraf, P. & H. Putnam (eds) 1983. *Philosophy of Mathematics: Selected Readings*, 2nd edn. Cambridge: Cambridge University Press.
Bishop, E. 1967. *Foundations of Constructive Analysis*. New York: McGraw-Hill.
Bishop, E. 1973. "Schizophrenia in Contemporary Mathematics", American Mathematical Society, Colloquium Lectures. Missoula: University of Montana.
Bishop, E. & D. Bridges 1985. *Constructive Analysis*. Heidelberg: Springer.
Bridges, D. 1998. "Constructive Truth in Practice". In *Truth in Mathematics*, H. Dales & G. Oliveri (eds). Oxford: Clarendon Press.
Boole, G. 2005. *An Investigation of the Laws of Thought*. New York: Barnes & Noble.
Boolos, G. 1998a. "Is Hume's Principle Analytic?". See Boolos (1998c), 301–14.
Boolos, G. 1998b. "The Standard of Equality of Numbers". See Boolos (1998c), 214–15.
Boolos, G. 1998c. *Logic, Logic and Logic*. Cambridge, MA: Harvard University Press.
Brouwer, L. E. J. 1954–56. "The Effect of Intuitionism on Classical Algebra of Logic". *Proceedings of the Royal Irish Academy*, vol. 57, Section A: Mathematical Astronomical, and Physical Science, 113–16. Dublin: Hodges, Figgis & Co.
Brouwer, L. E. J. 1983a. "Consciousness, Philosophy and Mathematics". See Benacerraf & Putnam (eds) (1983), 90–96.
Brouwer, L. E. J. 1983b. "Intuitionism and Formalism". See Benacerraf & Putnam (eds) (1983), 77–89.
Brown, J. R. 1999. *Philosophy of Mathematics: An Introduction to the World of Proofs and Pictures*. London: Routledge.
Cerbone, D. R. 2006. *Understanding Phenomenology*. Chesham: Acumen.
Church, A. 1996. *Introduction to Mathematical Logic*. Princeton, NJ: Princeton University Press.
Coffa, A. 1991. *The Semantic Tradition from Kant to Carnap: To the Vienna Station*. Cambridge: Cambridge University Press.

Colyvan, M. 2001. *The Indispensability of Mathematics*. Oxford: Oxford University Press.
Corfield, D. 2003. *Towards a Philosophy of Real Mathematics*. Cambridge: Cambridge University Press.
Curry, H. 1951. *Outline of a Formalist Philosophy of Mathematics*. Amsterdam: North Holland.
Curry, H. 1963. *Foundations of Mathematical Logic*. New York: McGraw-Hill.
Curry, H. 1983. "Remarks on the Definition and Nature of Mathematics". See Benacerraf & Putnam (eds) (1983), 202–6.
Dales, H. & G. Oliveri (eds) 1998. *Truth in Mathematics*. Oxford: Clarendon Press.
da Silva, J. J. 2000. Book Review, *Husserl or Frege? Meaning Objectivity and Mathematics* (Claire Ortiz Hill and Guillermo E. Rosado Haddock). *Manuscrito: Revista Internacional de Filosofia* **23**(2) (October), 351–71.
Dedekind, R. 1963. *Essays on the Theory of Numbers*. New York: Dover.
Degen, J. W. 1993. "Two Formal Vindications of Logicism". In *Philosophie der Mathematik*, J. Czermak (ed), *Scriftenreihe der Wittgensteingesellschaft* **20**(10), 243–50.
Detlefsen, M. 1986. *Hilbert's Program*. Dordrecht: Kluwer.
Dimitrov, R. 2002. *Computably Enumerable Vector Spaces, Dependence Relations and Turing Degrees*. PhD thesis submitted to the Department of Mathematics, George Washington University, Washington, DC.
Dummett, M. 1980. "Frege and Kant on Geometry". *Inquiry* **25**, 233–54.
Dummett, M. 1991a. *Frege, Philosophy of Mathematics*. London: Duckworth.
Dummett, M. 1991b. *The Logical Basis of Metaphysics*. Cambridge, MA: Harvard University Press.
Dummett, M. 2000. *Elements of Intuitionism*, 2nd edn. Oxford: Oxford University Press.
Field, H. 1980. *Science without Numbers: A Defence of Nominalism*. Oxford: Blackwell.
Field, H. 1989. *Realism, Mathematics and Modality*. Oxford: Blackwell.
Fraenkel, A. 1968. *Abstract Set Theory*. Amsterdam: North Holland.
Frege, G. 1952. "Frege on Russell's Paradox (appendix to Vol. II)". In *Translations from the Philosophical Writings of Gottlob Frege*, P. Geach & M. Black (eds and trans.). Oxford: Blackwell.
Frege, G. [1879] 1976. *Begriffsschrift, a Formula Language, Modeled upon that of Arithmetic, for Pure Thought*. In *From Frege to Gödel: A Source Book in Mathematical Logic, 1879–1931*, J. van Heijenoort (ed.), 1–82. Cambridge, MA: Harvard University Press.
Frege, G. 1979a. "On Euclidean Geometry". In G. Frege, *Posthumous Writings*, 167–9. Oxford: Blackwell.
Frege, G. 1979b. "A New Attempt at a Foundation for Arithmetic". In G. Frege, *Posthumous Writings*, 278–81. Oxford: Blackwell.
Frege, G. 1980. *Grundlagen*, 2nd ed. rev. Evanston, IL: Northwestern University Press.
Frege, G. [1884] 1980a. [*Die Grundlagen der Arithmetik*] *The Foundations of Arithmetic*, 2nd rev. edn, J. L. Austin (trans.). Evanston, IL: Northwestern University Press.
Frege, G. [1893, 1903] 1980b. [*Grundgesetze der Arithmetik*], partially translated in *Translations from the Philosophical Writings of Gottlob Frege*, 3rd edn, P. Geach & M. Black (eds and trans.), 117–224. Oxford: Blackwell.
Friend, M. 2006. "Meinongian Structuralism". In *The Logica Yearbook 2005*, M. Bílková & O. Tomala (eds). Prague: Filosofia.
Friend, M. n.d. "A *Reductio Ad Absurdum* Argument for Naïve Logicism". Unpublished manuscript.

Galilei, G. 1939. *Dialogues Concerning Two New Sciences*, H. Crew & A. Salvio (trans.). Evanston, IL: Northwestern University Press.
George, A. (ed.) 1994. *Mathematics and Mind*. Oxford: Oxford University Press.
George, A. & D. J. Vellman 2002. *Philosophies of Mathematics*. Oxford: Blackwell.
Giaquinto, M. 2002. *The Search for Certainty: A Philosophical Account of the Foundations of Mathematics*. Oxford: Clarendon Press.
Goble, L. 2001. *The Blackwell Guide to Philosophical Logic*. Oxford: Blackwell.
Gödel, K. [1944] 1983a. "Russell's Mathematical Logic". See Benacerraf & Putnam (eds) (1983), 447–69.
Gödel, K. [1964] 1983b. "What is Cantor's Continuum Problem?". See Benacerraf & Putnam (eds) (1983), 483–4.
Goethe, N. B. 2001. "Frege Between Kant and Leibniz or How to Understand Truth by Means of Rigorous Proof ". Manuscript presented to the History of Philosophy of Science Working Group (HOPOS), June 2002.
Hale, B. 1987. *Abstract Objects*. Oxford: Blackwell.
Hale, B. & C. Wright 2001a. "To Bury Caesar ...". In *The Reason's Proper Study: Essays Towards a Neo-Fregean Philosophy of Mathematics*, B. Hale & C. Wright (eds), 335–96. Oxford: Oxford University Press.
Hale, B. & C. Wright (eds) 2001b. *The Reason's Proper Study: Essays Towards a Neo-Fregean Philosophy of Mathematics*. Oxford: Clarendon Press.
Hallett, M. 1984. *Cantorian Set Theory and Limitation of Size*. Oxford: Clarendon Press.
Hallett, M. 1989. "Physicalism, Redutionism and Hilbert", in *Physicalism in Mathematics*, A. D. Irvine (ed.), 182–256. Dordrecht: Kluwer.
Hallett, M. 1991. "Hilbert, David". In *Handbook of Metaphysics and Ontology*, H. Burkhardt & B. Smith (eds), 354–8. Munich: Philosophia.
Hallett, M. 1994. "Hilbert's Axiomatic Method and the Laws of Thought". See George (ed.) (1994), 158–200.
Hardy, G. H. 1967. *A Mathematician's Apology*. Cambridge: Cambridge University Press.
Hart, W. D. (ed.) 1996. *The Philosophy of Mathematics*. Oxford: Oxford University Press.
Heath, P. L. 1967. "Nothing". In *The Encyclopedia of Philosophy*, vol. 5, P. Edwards (ed.), 524–5. New York: Macmillan.
Heck, R. Jr (ed.) 1997. *Language, Thought, and Logic*. Oxford: Oxford University Press.
Hellman, G. 1989. *Mathematics Without Numbers: Towards a Modal-Structural Interpretation*. Oxford: Clarendon Press.
Hellman, G. 2005. "Structuralism". In *The Oxford Handbook of Philosophy of Mathematics and Logic*, S. Shapiro (ed.), 556–62. Oxford: Oxford University Press.
Hilbert, D. 1922. "Neubegründung der Mathematik: Erste Mitteilung". *Abhandlungen aus dem mathematischen Seminar der Hamburgischen Universität* **I**, 157–77.
Hilbert, D. 1925. "Über das Unendliche", *Mathematische Annalen* **95**, 161–90.
Hilbert, D. 1971. *Foundations of Geometry*, 2nd edn, L. Unger (trans). La Salle, IL: Open Court.
Hill, C. O. 2000. "Frege's Attack on Husserl". In *Husserl or Frege? Meaning, Objectivity and Mathematics*, C. O. Hill & G. E. R. Haddock (eds), 95–108. Chicago, IL: Open Court.
Hill, C. O. & G. E. R. Haddock (eds) 2000. *Husserl or Frege? Meaning, Objectivity and Mathematics*. Chicago, IL: Open Court.
Isaacson, D. 1994. "Mathematical Intuition and Objectivity". See George (ed.) (1994), 118–40.
Kampis, G., L. Kvasz & M. Stöltzner (eds) 2002. *Appraising Lakatos: Mathematics, Methodology and the Man*. Dordrecht: Kluwer.

Kleene, S. C. 1971. *Introduction to Meta-Mathematics*. Amsterdam: North Holland.
Kneal, M. & W. Kneal 1962. *The Development of Logic*. Oxford: Oxford University Press.
Köhler, E. 2000. "Logic is Objective and Subjective". Paper presented at the History of Philosophy of Science Working Group (HOPOS) Conference, Vienna, 6–9 July.
Köhler, E. 2001. "Gödel on Intuition, and How Carnap Abandoned Empiricism". Unpublished manuscript.
Körner, S. 1962. *The Philosophy of Mathematics: An Introductory Essay*. New York: Harper & Row.
Lakatos, I. 1997. *Proofs and Refutations: The Logic of Mathematical Discovery*, J. Worrall & E. Zahar (eds). Cambridge: Cambridge University Press.
Larvor, B. 1998. *Lakatos: An Introduction*. London, Routledge.
Lemmon, E. J. 1965. *Beginning Logic*. Indianapolis, IN: Hackett.
Lucas, J. R. 2000. *The Conceptual Roots of Mathematics*. London, Routledge.
Machover, M. 1996. *Set Theory, Logic and their Limitations*. Cambridge: Cambridge University Press.
Maddy, P. 1990. *Realism in Mathematics*. Oxford: Clarendon Press.
Maddy, P. 1996. "Perception and Mathematical Intuition". In *The Philosophy of Mathematics*, W. D. Hart (ed.), 114–41. Oxford: Oxford University Press.
Maddy, P. 1997. *Naturalism in Mathematics*. Oxford: Clarendon Press.
Malinowski, G. 2001. "Many-Valued Logics". In *The Blackwell Guide to Philosophical Logic*, L. Goble (ed.), 309–35. Oxford: Blackwell.
Mancosu, P. 1998. *From Brouwer to Hilbert: The Debate on the Foundations of Mathematics in the 1920s*. Oxford: Oxford University Press.
Mill, J. S. 1970. *A System of Logic, Ratiocinative and Inductive*. London: Longman.
Moore, A. W. 1990. *The Infinite*. London: Routledge.
Musil, R. 1953. *The Man Without Qualities*. New York: Coward McCann.
Parsons, C. 1994. "Intuitionism and Number". See George (ed.) (1994), 141–57.
Paseau, A. 2005. "Naturalism in Philosophy and the Authority of Philosophy". *British Journal of the Philosophy of Science* **56**, 377–96.
Plato 2000. *The Republic*, G. R. Ferrari (ed.), T. Griffith (trans.). Cambridge: Cambridge University Press.
Potter, M. 2000. *Reason's Nearest Kin: Philosophies of Arithmetic from Kant to Carnap*. Oxford: Oxford University Press.
Potter, M. 2004. *Set Theory and its Philosophy*. Oxford: Oxford University Press.
Priest, G. 1987. *In Contradiction*. Dordrecht: Kluwer.
Priest, G. 2000. "Objects of Thought". *Australasian Journal of Philosophy* **78**, 494–502.
Priest, G. 2001. *An Introduction to Non-Classical Logic*. Cambridge: Cambridge University Press.
Priest, G. 2003. "Meinongianism and the Philosophy of Mathematics". *Philosophia Mathematica* **11** (February), 3–15.
Read, S. 1994. *Thinking About Logic: An Introduction to the Philosophy of Logic*. Oxford: Oxford University Press.
Resnik, M. 1981. "Mathematics as a Science of Patterns: Ontology and Reference". *Noûs* **15**, 529–50.
Resnik, M. 1982. "Mathematics as a Science of Patterns: Epistemology". *Noûs* **16**, 95–105.
Resnik, M. 1997. *Mathematics as a Science of Patterns*. Oxford: Clarendon Press.
Routley, R. 1980. *Exploring Meinong's Jungle and Beyond*. Canberra: RSSS, Australian National University.

Russell, B. 1903. *The Principles of Mathematics*. Cambridge: Cambridge University Press.
Russell, B. 1905. "Review of A. Meinong, Untersuchungen zur Gegenstadstheorie und Psychologie". *Mind* **14**, 530–38. Reprinted in B. Russell, *Essays in Analysis*, D. Lackey (ed.) (London: Allen & Unwin, 1973), 77–88.
Russell, B. 1907. "Review of: A. Meinong, Uber die Stellung der Gegestandtheorie im System der Wissenschaften". *Mind* **16**, 436–9. Reprinted in B. Russell, *Essays in Analysis*, D. Lackey (ed.) (London: Allen & Unwin, 1973), 89–93.
Russell, B. 1919. *Introduction to Mathematical Philosophy*. London: Allen & Unwin.
Sainsbury, R. M. 1988. *Paradoxes*. Cambridge: Cambridge University Press.
Shapiro, S. 1991. *Foundations Without Foundationalism*. Oxford: Clarendon Press.
Shapiro, S. 1997. *Philosophy of Mathematics: Structure and Ontology*. Oxford: Oxford University Press.
Shapiro, S. 2000. *Thinking About Mathematics*. Oxford: Oxford University Press.
Shapiro, S. (ed.) 2005. *The Oxford Handbook of Philosophy of Mathematics and Logic*. Oxford: Oxford University Press.
Shoenfield, J. R. 1994. *Mathematical Logic*. Nantick, MA: A. K. Peters.
Sieg, W. 1994. "Mechanical Procedures and Mathematical Experience". See George (ed.) (1994), 71–117.
Skorupski, J. 2005. "Later Empiricism and Logical Positivism". In *The Oxford Handbook of Philosophy of Mathematics*, 51–74. Oxford: Oxford University Press.
Sluga, H. 1977. "Frege's Alleged Realism". *Inquiry* **20**, 227–42.
Smullyan, R. 1990. *What is the Name of this Book? The Riddle of Dracula and Other Logical Puzzles*. Harmondsworth: Penguin.
Sylvan, R. 2003. "The Importance of Nonexistent Objects and of Intensionality in Mathematics". *Philosophia Mathematica* **11**, 20–52.
Tait, W. W. 1994. "The Law of Excluded Middle and the Axiom of Choice". See George (ed.) (1994), 45–70.
Tait, W. 2005a. "Finitism". See Tait (2005d), 21–43.
Tait, W. 2005b. "Remarks on Finitism". See Tait (2005d), 43–54.
Tait, W. 2005c. "The Law of Excluded Middle and the Axiom of Choice". See Tait (2005d), 105–32.
Tait, W. 2005d. *The Provenance of Pure Reason: Essays in the Philosophy of Mathematics and its History*. Oxford: Oxford University Press.
Tennant, N. 1987. *Anti-Realism and Logic: Truth as Eternal*. Oxford: Oxford University Press.
Tennant, N. 1992. *Autologic*. Edinburgh: Edinburgh University Press.
Tennant, N. 1997. *The Taming of the True*. Oxford: Clarendon Press.
Tieszen, R. 2005. *Phenomenology, Logic and the Philosophy of Mathematics*. Cambridge: Cambridge University Press.
Tomassi, P. 1999. *Logic*. London: Routledge.
Van Heijenoort, J. (ed.) 1977. *From Frege to Gödel: Mathematical Logic, 1879–1931*. Cambridge, MA: Harvard University Press.
Van Stigt, W. P. 1998. "Brouwer's Intuitionist Programme". In *From Brouwer to Hilbert: The Debate on the Foundations of Mathematics in the 1920s*, Paolo Mancosu (ed.), 1–22. Oxford, Oxford University Press,.
Vlastos, G. 1967. "Zeno of Elea". In *The Encyclopedia of Philosophy*, vol. 8, 369–79. New York: Macmillan.

Weir, A. 1998. "Naïve Set Theory is Innocent!". Mind **107**, 763–98.
Whitehead, A. N. & B. Russell 1910–13. *Principia Mathematica*, 3 vols. Cambridge: Cambridge University Press.
Wright, C. 1983. *Frege's Conception of Numbers as Objects*. Aberdeen: Aberdeen University Press.
Zalta, E. N. 1988. *Intensional Logic and the Metaphysics of Intentionality*. Cambridge, MA: MIT Press.

Index

abstraction
   principles 75; *see also* Hume's principle, context principle, numbers principle, basic law V
   process of 91
Achilles 3–6, 12, 177
acquisition argument 117–19
actual world 86, 173, 158–9, 161
actual infinity *see* infinity, actual
algorithmic learning theory 129, 152
analysis 33, 50, 57, 78–9, 136, 183
analytic truth ix, 60–63, 73, 75; *see also* synthetic truth
ancient Egypt 95
ancient Greece 1, 26, 31, 52
ante rem 90
anti-realist 31–2, 35, 40, 49–50, 98, 101, 103–4, 153–4, 183, 187
a posteriori 61
a priori 23, 29, 60, 73, 132
Aristotle 2–3, 7–9, 11, 52, 177, 188
arithmetic *see* geometry, non-standard models of, axioms: Peano/Dedekind, quantifier-free elementary
Austin, J. L. 58
axioms
   choice 42–3, 97, 102, 106, 120, 188
   empirical 70–71
   Euclidean geometry 24, 46; *see also* geometry, Euclidean
   independent 42
   induction 56–7, 72
   infinity 41–3
   logic 60; *see also* basic laws
   logically necessary 70–71
   Peano/Dedekind of arithmetic 52, 55, 84, 86–8
   scheme 57
   set-theoretic 27, 31–2, 42–3, 97; *see also* Zermelo–Fraenkel
   type theory 67–8

Babylonian mathematics 95, 186
basic law 60, 67–8, 71–3, 75, 78, 187
   V 63–6, 71–2, 75, 156, 175
Begriffsschrift (Frege) 49, 71, 181
Beltrami, Eugenio 24
Benacerraf, Paul 83, 85–6, 89, 94, 97, 99, 184–5
Bishop, Errett 123, 188
bivalence 33, 106, 108–9, 121
Bolyai, Jànos 24
Boole, George 52–4
Boolos, George 74–7, 181
Bridges, Douglas 123, 188
Brouwer, L. E. J. 101, 114, 116–17, 120, 123, 154
Burali-Forti paradox 26, 113–14, 171

calculus 3, 33, 169
Cantor, Georg 34–5, 57, 71, 120, 145, 154, 156, 178
   diagonal argument 2, 20, 34, 57, 113
   paradise of 2, 31, 120, 154
   paradox 113–14
cardinality 14–16, 18–21, 39, 56–7, 98, 113, 156, 186–7
   same 14, 17, 21, 56
category theory 29
causation 37–40, 44, 46–7, 103, 128, 143, 159, 180–81
Clarke, Samuel 53

class 27–8, 31, 42, 45, 76, 98, 104, 137, 179
Coffa, Alberto 63
compact 95, 182, 186
compositionality 122–3
computer 115–16, 129, 145, 149, 151–2, 166, 189
  science 129
  scientist 67, 69, 129, 148
consistency 66, 94, 148, 150–51, 155, 157, 180, 184, 189
  equi-consistency 180, 184
context principle 74
continuum problem 21–2, 35, 179
Corfield, David 166
counterfactual 158–9

decidable 95
Dedekind, Richard 16, 49, 52, 55, 84, 86–8
density of numbers see rational numbers
discharging rule 111–12, 187
disjunctive syllogism 110–11, 113, 167, 187
double negation 102, 106–8, 111–12, 117, 119–20, 167
Dummett, M. 117, 120–22, 124

economics 51, 59
effective see procedure
eliminativist 82, 97, 138
empiricism 1, 45, 51, 127, 129–33, 142–4, 147
equivalence class 98
ex falso quod libet 107–8, 114, 159–62, 167–8
existential 65, 160
  proof 102, 115–16
  quantifier 33–4, 160, 183
expressive power 53–4, 66, 69, 100
extension of a concept 63–6

fictionalism 51, 127, 134
Field, Hartry 134–6
formalism ix, 127–9, 147–9, 151–2, 154, 165, 178
forms see Platonic forms

gapless
  line 21–2, 169
  proof 60–61
geometry

Euclidean 24–5, 30, 32, 46, 51–2, 59, 66, 161, 164
  non-Euclidean 25, 92, 164
  parallel postulate 24–5
Gilgamesh Epic 2
Gödel, Kurt 23, 35–7, 39, 44, 45, 155, 157
Gödel–Bernays set theory 43, 43
group theory 33

Hale, B. 72, 76–7, 79
Hellman, Geoffrey 81–2, 85–92, 96–7, 138, 158, 162
Heyting, Arend 120
Hilbert, David 1–2, 13–14, 123–4, 127, 153–7, 188–90
Hume, David 74–6
Husserl, Edmund 1, 127, 130, 137, 141–7, 153, 163

idealism, Kantian 104
Ideals see Platonic forms
impredicative definitions 35–6
in re 90, 93, 96
indispensability arguments 128, 135–7
integers 16–17, 19, 93, 96
intension 64, 183
irrational numbers 19, 177–8
  decimal expansion of 19

Julius Caesar problem 66, 72

Kant 29, 62
Köhler 44–7, 76–7
Kronecker, Leopold 123
Lakatos, Imre 127, 130, 163–6
law of excluded middle 33–4, 102, 106, 108–9, 112, 117, 120–21, 123
Leibniz 158, 182
Lobachevsky, Nikolai Ivanovich 24
logic
  constructive 36, 102, 105, 116, 124
  free 77
  higher-order 34
  law of 64–5, 72, 120, 156
  modal 77, 86
  non-classical 33–4, 78
  paraconsistent 160–62
  propositional 53–4, 78, 95, 108, 123, 181–2, 188
  relevant 113, 160

second-order 54, 57, 62, 75–6, 78, 162–3, 182, 185
soundness of 108, 187
temporal 77
Löwenheim–Skolem property 95, 182, 186

Maddy, Penelope 23, 29, 36–40, 43–7, 121, 180
manifestation argument 117–20
Meno (Plato) 24
Meinong, Alexius 127, 130, 157–9, 161–3, 190
Mill, J. S. 128, 130–31
model theory ix, 29, 94–6, 99–100
modus ponens 61, 73, 111, 182
modus tollens 106, 111, 167
Moosbrugger 134–5, 189

natural deduction 109–10, 121, 148, 187, 189
naturalism 1, 40, 45, 51, 127–8, 130, 133
neuroscience 138–9, 143
noetic experience 143
non-standard models of arithmetic 57, 182
normativity 11, 97, 114, 116, 134, 141, 184
noumenal world 104–5

object
  concrete 158
  contradictory 158–9, 190
  logical 49, 58–60, 72
  perfect 25, 29–30, 46
  real 59, 91
  spooky 9, 89, 92, 97
Ockham 41
one-to-one correspondence 15–18, 20–21, 74–5, 182
ordering relation 12
ordinal
  construction 42
  numbers 12–13, 22, 85
  limit 13

parities principle 75–7
Parmenides 3
Peano, Giuseppe 60, 71–2, 179; *see also* Dedekind
Peano/Dedekind axioms 52, 55, 84, 86–8
perception 30–31, 34–5, 37–40, 43–5, 105, 131, 135, 143, 155

phenomenal world 104–5
Plato, Forms 25, 29–31
Potter, Michael 147
powerset 21–2, 32, 41, 113, 120
premises 106–9, 112, 123, 160, 187–8
prescriptivity 11, 77, 97, 116
Priest, Graham 159–62
prime pairs 105, 103, 187
Principia Mathematica 49
Pythagoreans 145
Pythagoras 24–5

quantifier-free elementary arithmetic 147
Quine, W. V. 37

reductio ad absurdum 33–4, 102, 106–8, 116–17, 120, 167, 181, 188
Resnik, Michael 81, 83, 85, 90–91, 96
rigour 32, 47, 62, 134–5, 153–4
Routley, Richard 158–60, 162
Russell, Bertrand *see* Whitehead, Alfred North
  paradox 113–14

second-order logic *see* logic, second-order
set *see* class
  construction 27, 29–31, 36
  empty 15, 27, 32–4, 41–2, 65, 96, 180, 183, 184–5
  mathematical objects 12, 36, 38, 83, 86, 136, 180, 185
  size of 14–15, 31, 34, 40–41, 71, 95, 120, 157
  subset of 15–16, 20–21, 83, 93–4, 156, 178
Shapiro, Stewart 37, 81, 83, 85, 90, 93, 95–6, 98–9, 154
singular term 58
Socrates 24–5, 179
Spinoza 53
Stricker 130, 137
structure
  free-standing 95
  meta-structure 87–8, 95–6, 98–9
  sui generis 90
successor 13, 83–4, 56, 86
  immediate 13, 55–6, 94, 182
supervenience 39–40, 44, 92, 129, 180, 185
syntax 33, 65, 108–9, 155, 183, 187–8
synthetic truth x, 60–62, 73, 76–7, 184

tautology 11, 59–60, 112, 187
Tennant, Neil 117–18, 181, 184
thermodynamics, second law of 10
topology 25, 29, 50, 136
tortoise 4–6, 12, 177
truth-apt 103–5, 109
Turing test 151–4
type theory *see* Whitehead

universal quantifier 55, 65, 69, 74–5, 78, 122, 165–6

validity 53–4, 65, 106–7, 112, 167–8, 186

Weyl, Hermann 123
Whitehead, Alfred North 49–52, 66–71, 78–9, 183
Wittgenstein xi, 37, 163
Wright, Crispin 71–7, 79

Zeno 1–3, 7, 114, 177
Zermelo, Ernst 21–2, 42–3, 83, 88–9, 99, 114, 178, 183, 184
Zermelo–Fraenkel set theory *see* Zermelo